Topics in LC Oscillators

Konstantinos Manetakis

Topics in LC Oscillators

Principles, Phase Noise, Pulling, Inductor
Design

 Springer

Konstantinos Manetakis
Integrated & Wireless Systems
CSEM Neuchâtel
Neuchâtel, Switzerland

ISBN 978-3-031-31088-1   ISBN 978-3-031-31086-7   (eBook)
https://doi.org/10.1007/978-3-031-31086-7

This Springer imprint is published by the registered company Springer Nature Switzerland AG
The registered company address is: Gewerbestrasse 11, 6330 Cham, Switzerland

To my family

Preface

This book introduces what the author believes to be some of the fascinating topics in LC oscillators. These involve self-sustained oscillators, oscillator noise, oscillator entrainment, pulling, and the related subjects of magnetic coupling and integrated inductor design.

The book derives from the author's effort to comprehend some of the topics covered. As such, it constitutes a fresh approach to the subject, and it thus may be helpful to electrical engineering students and practicing engineers. The presentation emphasizes the basic principles and illuminates them with approximate calculations. Such an approach imparts intuition and design insight and serves as a complement to the use of simulators.

In Chap. 1, we review the basics of linear oscillators. We build upon the fundamental model of the simple harmonic oscillator, introduce the concept of the phase plane, and extend the discussion to cover the damped harmonic oscillator and its energy. We subsequently present linear oscillator models and underline their shortcomings in describing practical oscillators that are inherently nonlinear.

In Chap. 2, we discuss self-sustained oscillators and highlight their primary future of possessing their own rhythm and oscillation amplitude, defined entirely by their internal properties. We introduce the paradigmatic self-sustained Van der Pol oscillator model, formulate the Van der Pol equation, and develop methods to obtain its solution. We close the chapter by demonstrating that the Vad der Pol oscillator can be used to model realistic oscillators and accurately predict the amplitude and frequency of oscillation.

Chapter 3 is an introduction to oscillator noise. We establish two different viewpoints connected by the fluctuation-dissipation theorem. We utilize both approaches to analyze stationary noise and derive Leeson's formula for phase noise. In a related appendix, we apply the fluctuation approach to the case of amplitude noise and demonstrate that phase noise dominates oscillator noise.

In Chap. 4, we present the phase dynamics equation that governs the time evolution of the oscillator's phase in the presence of disturbances. It models an oscillator as a phase point moving along the oscillator's limit cycle, occasionally disturbed by noise. This approach describes the noise to phase noise conversion as a two-step process: noise translation from the oscillator harmonics to the oscillator carrier frequency and, subsequently, the conversion of the translated noise to phase noise. We demonstrate the effectiveness of this decomposition by applying it to model cyclo-stationary noise and accurately capture transconductor thermal noise.

Chapter 5 applies the phase dynamics equation to the case of low-frequency cyclostationary noise sources. We focus on transconductor flicker noise and supply/bias low-frequency noise and obtain expressions for the phase noise. Based on the analysis, we identify the vital role of the oscillator's common-mode behavior in controlling the impact of low-frequency noise on phase noise. We close the chapter by discussing phase noise due to varactors.

In Chap. 6, we concentrate on oscillator entrainment. We demonstrate the versatility of the phase dynamics equation to describe injection locking, predict the locking range, and accurately model the oscillator's complex motion when the injection signal is not strong enough to entrain it. We also utilize the phase dynamics equation to formulate Adler's model for oscillator pulling. The rest of the chapter focuses on magnetic coupling and practical techniques to alleviate its influence.

In the final Chap. 7, we focus on the analysis and design of integrated inductors. We discuss methods to accurately estimate the inductance value and account for magnetic and electric loss mechanisms that hinder inductor performance. We build a simple inductor equivalent network to dimension the inductor structure before resorting to electromagnetic simulations.

Neuchâtel, Switzerland   Konstantinos Manetakis

Contents

Chapter 1
Basics of LC Oscillators

1.1 Introduction

This chapter presents the most basic oscillator model, the simple harmonic oscillator. We introduce the concept of the phase plane and extend our discussion to the damped and driven simple harmonic oscillator models. We consider the relationship of the quality factor to the rate of change of the oscillator energy and the oscillator lineshape. We finally present linear oscillator models and highlight their usefulness and limitations.

1.2 The Simple Harmonic Oscillator

The most fundamental oscillator model is the simple harmonic oscillator. This model applies to diverse physical systems such as the spring-mass system, the simple pendulum, the air mass in the neck of an open flask, an electric circuit containing capacitance and inductance, etc. The state of such systems when slightly perturbed from their equilibrium or rest position can be described under appropriate simplifications by the equation of simple harmonic motion [1].

Let us assume that at time $t = 0$, we connect an inductor L across a capacitor C, which is initially charged to voltage V_o as is shown in Fig. 1.1. Oscillatory behavior is explained intuitively as follows. The charge concentration on the capacitor's top plate imparts a flow of current through the inductor. The current is maximum when the charge is equally distributed on the two plates. The inductor opposes the change in the current flow through it, and the current continues to flow. More charge is thus collecting on the opposite capacitor plate. When the current flow becomes zero, all the charge crowds on the opposite capacitor plate, and the process reverses. In the absence of any loss mechanism, the cycle repeats in perpetuity [1].

The current out of the capacitor equals the current into the inductor. Therefore

K. Manetakis, *Topics in LC Oscillators*,
https://doi.org/10.1007/978-3-031-31086-7_1

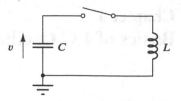

Fig. 1.1 Lossless LC tank circuit. The capacitor is charged to V_o and the switch closes at time $t = 0$

$$C\frac{dv}{dt} + \frac{1}{L}\int v\,dt = 0 \tag{1.1}$$

where v is the voltage across the tank. Differentiating with respect to time gives

$$\ddot{v} + \omega_o{}^2 v = 0 \tag{1.2}$$

where over-dot denotes time derivative. The constant ω_o is the angular frequency given by

$$\omega_o = \sqrt{\frac{1}{LC}} \tag{1.3}$$

(1.2) is the differential equation of simple harmonic motion. We look for solutions of the form $v \propto e^{\lambda t}$ where λ is complex [2]. The imaginary part of λ captures the oscillatory behavior, while the real part captures growth or decay with time. Substituting this last expression in (1.2) gives the characteristic equation

$$\lambda^2 + \omega_o{}^2 = 0 \tag{1.4}$$

which has solutions $\lambda_{1,2} = \pm j\omega_o$. The exponent λ is imaginary in accordance with the lack of any energy decay mechanism in Fig. 1.1. The general solution of (1.2) can be written as the linear combination [2]

$$v = Ae^{j\omega_o t} + Be^{-j\omega_o t} \tag{1.5}$$

where A, B are arbitrary constants that can be determined by the initial conditions: $v(0) = V_o$ and $\dot{v}(0) = 0$ that apply to the experiment shown in Fig. 1.1. We thus obtain the system of equations

$$V_o = A + B \tag{1.6}$$

$$0 = j\omega_o A - j\omega_o B \tag{1.7}$$

from which the solution of (1.2) can be written as

$$v = V_o \cos(\omega_o t) \tag{1.8}$$

Fig. 1.2 Time evolution of
lossless LC tank
($C = 1$ pF, $L = 1$ nH) with
initial conditions $v(0) = 1$ V
and $\dot{v}(0) = 0$

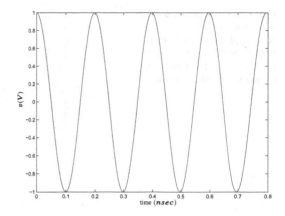

This is a harmonic oscillation with cyclic frequency $f_o = \omega_o/2\pi$ cycles per unit
time, as shown in Fig. 1.2.

1.3   Phase Plane

To describe the state of the lossless LC tank circuit shown in Fig. 1.1 at some instant
in time, it is not enough to know the voltage v across the capacitor. Indeed, for
the same value of v, the capacitor voltage may be increasing or decreasing. We,
therefore, need one more variable. To formally introduce it, we note that the second-
order differential equation of simple harmonic motion (1.2) can be written as two
coupled first-order differential equations [3]

$$\dot{x} = y \tag{1.9}$$

$$\dot{y} = -\omega_o^2 x \tag{1.10}$$

The state of the lossless LC tank circuit is specified at any instant in time by the
voltage $v = x$ across the capacitor, together with its derivative $\dot{v} = y$ at that instant,
and can be represented by a point on a two-dimensional phase space (phase plane)
with coordinates (x, y). Knowing the time derivative of the capacitor voltage is
equivalent to knowing the current through the capacitor since $i_C = C\dot{v}$. A physical
system characterized by equations such as (1.9) and (1.10), where time does not
appear explicitly on the right side, is known as autonomous as opposed to driven
[4].

Over time, the system traverses a path on the phase plane. Since it undergoes an
oscillatory motion, we expect the phase path to close upon itself after a time equal to
the period of oscillation $T_o = 2\pi/\omega_o$. To mathematically describe the allowed phase
paths, we use the chain rule to eliminate time in (1.9), (1.10), and after integration,
we obtain

Fig. 1.3 Phase plane path for
a lossless LC tank
($C = 1$ pF, $L = 1$ nH) with
initial conditions $v(0) = 1V$
and $\dot{v}(0) = 0$. The path is an
ellipse traversed clockwise

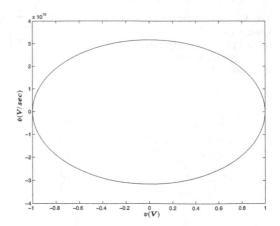

$$y^2 + \omega_o{}^2 x^2 = constant \tag{1.11}$$

This final equation describes a family of ellipses. Figure 1.3 shows an ellipse corresponding to initial capacitor voltage $V_o = 1$ V for $C = 1$ pF and $L = 1$ nH. The ellipse is traversed clockwise as the voltage across the capacitor increases with time for positive current and vice versa. By substituting $x = v = V_o \cos(\omega_o t)$ and $y = \dot{v} = -\omega_o V_o \sin(\omega_o t)$ into (1.11) we find that the value of the constant equals $(\omega_o V_o)^2$, where V_o is the peak voltage across the capacitor. Since the energy

$$E = \frac{1}{2} C V_o{}^2 \tag{1.12}$$

is conserved, the system, once set in motion, follows the same ellipse on the phase plane in perpetuity [9].

1.4   The Damped Harmonic Oscillator

In any real LC oscillator, energy dissipating mechanisms reduce the oscillation amplitude over time [1]. One way to model such effects is with a parallel resistor, as shown in Fig. 1.4.

Assuming that the capacitor is initially charged to voltage V_o and we close the switch at time $t = 0$, the equation for the damped-LC tank circuit becomes

$$\ddot{v} + \frac{1}{RC}\dot{v} + \frac{1}{LC}v = 0 \tag{1.13}$$

and in standard form [5]

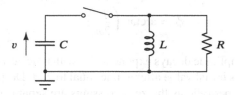

Fig. 1.4 Damped-LC tank circuit. The capacitor is charged to V_o, and the switch closes at time $t = 0$. Energy is dissipated in the resistor R, which is connected in parallel to the tank

$$\ddot{v} + \beta\dot{v} + {\omega_o}^2 v = 0 \qquad (1.14)$$

$\beta = 1/RC$ is the damping rate. Equation (1.14) is the differential equation of damped harmonic motion. We look for solutions of the form $v \propto e^{\lambda t}$. Substituting this last expression in (1.14) gives the characteristic equation

$$\lambda^2 + \beta\lambda + {\omega_o}^2 = 0 \qquad (1.15)$$

which for the under-damped case ($\omega_o > \beta/2$) has solutions

$$\lambda_{1,2} = -\frac{\beta}{2} \pm j\sqrt{{\omega_o}^2 - \frac{\beta^2}{4}} \qquad (1.16)$$

The general solution of (1.14) can thus be written as

$$v = Ae^{\lambda_1 t} + Be^{\lambda_2 t} \qquad (1.17)$$

where A, B are arbitrary constants that can be determined by the initial conditions: $v(0) = V_o$ and $\dot{v}(0) = -V_o/RC$. We thus obtain the system of equations

$$V_o = A + B \qquad (1.18)$$

$$-\frac{V_o}{RC} = A\lambda_1 + B\lambda_2 \qquad (1.19)$$

from which we find the solution of (1.14) as

$$v = \frac{V_o}{\cos\phi} e^{-\frac{\beta}{2}t} \cos(\omega t + \phi) = V_o e^{-\frac{\beta}{2}t} \left[\cos(\omega t) - \frac{\beta}{2\omega} \sin(\omega t) \right] \qquad (1.20)$$

where

$$\omega = \omega_o \sqrt{1 - \left(\frac{\beta}{2\omega_o}\right)^2} \qquad (1.21)$$

$$\phi = \arctan\left(\frac{\beta}{2\omega}\right) \tag{1.22}$$

The oscillation amplitude decays exponentially with time, as is shown in Fig. 1.5, and falls to 37% of its initial value after a time equal to $2/\beta$. Despite the exponential decay, the motion is periodic as the zero crossings are separated by time intervals equal to $2\pi/\omega$ [5]. Compared to the undamped case, (1.21) shows that the frequency is reduced due to the damping.

On the phase plane, the damped harmonic oscillator is represented by the pair of equations

$$\dot{x} = y \tag{1.23}$$

$$\dot{y} = -\beta y - \omega_o{}^2 x \tag{1.24}$$

which for positive β describe a stable, diminishing spiral as is shown in Fig. 1.6.

Fig. 1.5 Time response of damped-LC tank circuit ($L = 1$ nH, $C = 1$ pF, $R = 500\ \Omega$) with initial conditions $v(0) = V_o = 1$ V and $\dot{v}(0) = -V_o/RC$. The response is an exponentially decaying sinewave

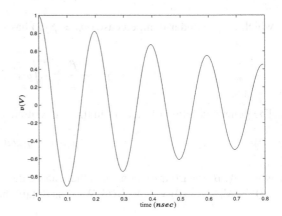

Fig. 1.6 Phase plane path for the damped resonant circuit ($C = 1$ pF, $L = 1$ nH, $R = 500\ \Omega$) with initial conditions $v(0) = V_o = 1$ V and $\dot{v}(0) = -V_o/RC$. Energy dissipation in the resistor R results in a stable, diminishing spiral

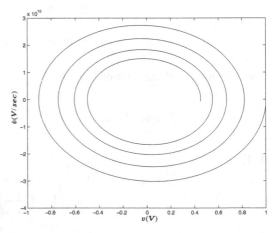

1.5 Energy in the Damped Harmonic Oscillator

Under the assumption of small loss: $\beta/2\omega_o \ll 1$, we have that $\omega \approx \omega_o$ and $\phi \approx 0$. Equation (1.20) can thus be approximated as

$$v = V_o e^{-\frac{\beta}{2}t} \cos(\omega_o t) \tag{1.25}$$

Furthermore, the inductor current can be approximated by $i_L = -C\dot{v}$, and the energy stored in the oscillator can be written as

$$E = \frac{1}{2}C\,v^2 + \frac{1}{2}L\,i_L{}^2 \tag{1.26}$$

Substituting (1.25) into (1.26) gives for the oscillator energy

$$E = \left(\frac{1}{2}CV_o{}^2\right) \cdot e^{-\beta t} = E_o \cdot e^{-\beta t} \tag{1.27}$$

where E_o is the initial oscillator energy. Under small loss conditions, the oscillator energy decreases exponentially with time, with a rate given by

$$\left|\frac{dE}{dt}\right| = \beta E \tag{1.28}$$

The oscillator energy falls to 37% of its initial value after a time equal to $\tau = 1/\beta$ [5].

To obtain a measure of the quality of the oscillator, we may compare time τ to the period of the oscillator T_o

$$\frac{\tau}{T_o} = \frac{1}{2\pi}\frac{\omega_o}{\beta} \tag{1.29}$$

The ratio ω_o/β is termed the tank quality factor Q [5]

$$Q = \frac{\omega_o}{\beta} = R\omega_o C = \frac{R}{\omega_o L} \tag{1.30}$$

Furthermore, over a time interval equal to the period T_o, we have

$$\frac{E(t + T_o)}{E(t)} = e^{-\beta T_o} \approx 1 - \beta T_o \tag{1.31}$$

where the approximation holds for adequately small losses. Therefore

$$\frac{E(t) - E(t + T_o)}{E(t)} = \frac{\Delta E}{E} \approx \beta T_o = \frac{2\pi}{Q} \tag{1.32}$$

where ΔE is the energy loss per oscillation cycle. The quality factor Q thus gives the oscillation energy divided by the energy loss per cycle per radian

$$Q = 2\pi \frac{E}{\Delta E} \tag{1.33}$$

The smaller the energy loss, the higher the oscillator quality factor.

1.6   The Driven Harmonic Oscillator

Figure 1.7 shows an under-damped-LC oscillator driven by a current source. The circuit equation becomes

$$i_{in} = C\frac{dv}{dt} + \frac{v}{R} + \frac{1}{L}\int v\,dt \tag{1.34}$$

For the driving current, we assume

$$i_{in} = i_o \sin(\omega t) \tag{1.35}$$

Differentiating Eq. (1.34) to remove the integral and substituting $\beta = 1/RC$ and $\omega_o{}^2 = 1/LC$ we obtain

$$\ddot{v} + \beta\dot{v} + \omega_o{}^2 v = \frac{i_o\omega}{C}\cos(\omega t) \tag{1.36}$$

When the driving current is applied, the oscillator is disturbed from its equilibrium position. It is inclined to oscillate at its natural frequency ω_o, but due to the applied current at frequency ω, the resultant motion is the superposition of two oscillations with frequencies ω_o and ω, respectively. The motion of the system during this initial period is called the transient response (see Fig. 1.8). We are interested in the steady-state solution of Eq. (1.36) that describes the motion of the system after the driving current has been applied for a long enough time so that

Fig. 1.7 Damped-LC tank circuit driven by an external current source

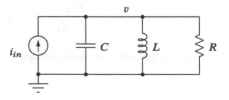

Fig. 1.8 Driven damped-LC tank circuit response with $L = 1$ nH, $C = 1$ pF, and $R = 500\Omega$. The driving current has 1 mA of amplitude and a frequency 5.3 GHz. Steady-state is attained after about 5 nsec

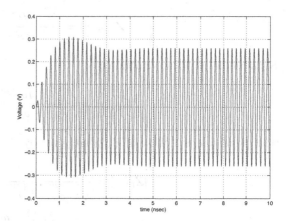

the transient response has decayed to zero. The steady-state solution thus describes the oscillator's motion after the driving current has had enough time to enforce its will. The oscillator then undergoes harmonic motion at the driving frequency ω with some phase difference with respect to the applied current. It is thus reasonable to assume that the steady-state solution of Eq. (1.36) is of the form [1]

$$v = v_o \cdot \sin(\omega t + \theta) \tag{1.37}$$

where v_o and θ are functions of the angular frequency.

In order to determine v_o and θ, we substitute (1.37) into (1.36). This results in the pair of equations

$$v_o \cdot \left[\beta \omega \cos(\theta) + (\omega_o^2 - \omega^2) \sin(\theta) \right] = \frac{i_o \, \omega}{C} \tag{1.38}$$

$$\beta \omega \sin(\theta) - (\omega_o^2 - \omega^2) \cos(\theta) = 0 \tag{1.39}$$

Equation (1.39) gives

$$\theta = \arctan \left[\frac{\omega_o^2 - \omega^2}{\beta \omega} \right] \tag{1.40}$$

Substituting (1.40) into (1.38) and using the identities $\cos[\arctan(x)] = 1/\sqrt{1 + x^2}$ and $\sin[\arctan(x)] = x/\sqrt{1 + x^2}$ results in

$$\frac{v_o}{i_o} = \frac{\omega/C}{\sqrt{(\omega_o^2 - \omega^2)^2 + (\beta \omega)^2}} \tag{1.41}$$

The physical interpretation of (1.41) and (1.40) becomes clear if we calculate the impedance looking into the tank

$$Z_{tank}(s) = \frac{s/C}{s^2 + \beta s + \omega_o^2} \tag{1.42}$$

Substituting $j\omega$ for s we obtain

$$Z_{tank}(j\omega) = \frac{j\omega/C}{(\omega_o^2 - \omega^2) + j\omega\beta} \tag{1.43}$$

The magnitude and phase of Z_{tank} are given by

$$|Z_{tank}| = \frac{\omega/C}{\sqrt{(\omega_o{}^2 - \omega^2)^2 + (\beta\omega)^2}} \tag{1.44}$$

$$\angle Z_{tank} = \arctan\left(\frac{\omega_o{}^2 - \omega^2}{\beta\omega}\right) \tag{1.45}$$

which are the same as (1.41) and (1.40), respectively.

The steady-state response of the tank to the driving current $i_{in} = i_o \sin(\omega t)$ can thus be expressed as the product

$$v = i_{in} \cdot Z_{tank} \tag{1.46}$$

which further becomes

$$v = i_o |Z_{tank}| \cdot \sin(\omega t + \angle Z_{tank}) \tag{1.47}$$

The amplitude of the driving current i_o times the magnitude of the tank impedance $|Z_{tank}|$ equals the magnitude of the response, while the phase of the response follows the phase of the tank impedance. The amplitude and the phase of the response versus the frequency of the applied driving current are plotted in Fig. 1.9. The voltage across the tank remains very small for frequencies sufficiently far away from the natural frequency of the tank ω_o. It obtains its maximum value $V_o = i_o R$ when the driving frequency ω equals ω_o, a condition known as resonance [5]. Furthermore, for frequencies $|\omega - \omega_o| \gg \beta$, the tank's response is 90 degrees out of phase with respect to the driving current.

1.7   Energy in the Driven Harmonic Oscillator

In the steady-state, the average energy stored in the electric and magnetic fields of the LC tank is

$$\overline{E_{stored}} = \overline{\frac{1}{2}C\,v^2} + \overline{\frac{1}{2}L\,(C\,\dot{v})^2} \tag{1.48}$$

Fig. 1.9 Steady-state
response of driven
damped-LC tank network
with $L = 1$ nH, $C = 1$ pF and
$R = 500\Omega$. Driving current
has amplitude $i_o = 1$ mA

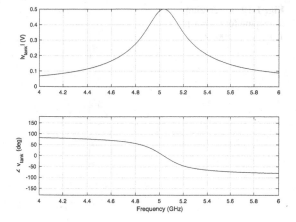

Plugging (1.47) in (1.48) results in

$$\overline{E_{stored}} = \frac{1}{4} C\, i_o{}^2 \,|Z_{tank}|^2 \left(1 + \frac{\omega^2}{\omega_o{}^2}\right) \tag{1.49}$$

Substituting for $|Z_{tank}|$ from (1.44) gives

$$\overline{E_{stored}} = \frac{1}{4} C\, i_o{}^2 \,\frac{\omega^2/C^2}{(\omega_o{}^2 - \omega^2)^2 + (\beta\omega)^2}\left(1 + \frac{\omega^2}{\omega_o{}^2}\right) \tag{1.50}$$

For driving frequencies very close to the natural frequency ω_o, there is not much error introduced by replacing ω by ω_o everywhere except in the term $(\omega_o^2 - \omega^2)^2$. This term can be simplified as $(\omega_o{}^2 - \omega^2) = (\omega_o + \omega)(\omega_o - \omega) \approx 2\omega_o(\omega_o - \omega)$ [5]. Equation (1.50) then becomes

$$\overline{E_{stored}} \approx \frac{1}{2} C\, V_o{}^2 \left[\frac{(\beta/2)^2}{(\omega_o - \omega)^2 + (\beta/2)^2}\right] \tag{1.51}$$

where $V_o = i_o R$ is the peak voltage at resonance. The term in the brackets is the Lorentzian lineshape function and is plotted in Fig. 1.10. It quantifies the frequency dependence of the oscillator's energy when the oscillator is excited by a periodic driving source [5]. The oscillator absorbs energy from the external source according to (1.51). The bandwidth BW, of the resonance curve, can be found by solving

$$|\omega - \omega_o| = \beta/2 \Rightarrow \omega = \omega_o \pm \beta/2 \tag{1.52}$$

and is therefore equal to the damping rate

$$BW = \beta \tag{1.53}$$

Fig. 1.10 Lorentzian
lineshape for $C = 1$ pF, $L = 1$ nH, $R = 500 \, \Omega$

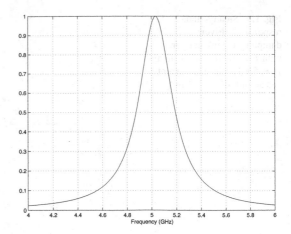

Frequency (GHz)

(1.30) and (1.53) allow us to express the quality factor Q as

$$Q = \frac{\omega_o}{\beta} = \frac{\omega_o}{BW} \qquad (1.54)$$

The frequency selectivity of the driven oscillator, that is, its response to different driving frequencies is characterized by its quality factor. The higher the Q, the narrower the resonance bandwidth.

The driving source provides energy that the oscillator absorbs. At resonance, the average energy stored in the oscillator can be found by substituting $\omega = \omega_o$ in (1.51)

$$\overline{E_{stored}}(\omega_o) = \frac{1}{2} C V_o^2 \qquad (1.55)$$

while the average energy per cycle that is dissipated in the resistor R is

$$\overline{E_{loss}}(\omega_o) = \frac{V_o^2}{2 R} T_o = \frac{V_o^2}{2 R} \frac{2\pi}{\omega_o} \qquad (1.56)$$

Using (1.56) and (1.55), we can show that the ratio of the average energy stored in the oscillator at resonance divided by the average energy loss per cycle per radian equals the quality factor

$$2\pi \frac{\overline{E_{stored}}(\omega_o)}{\overline{E_{loss}}(\omega_o)} = \frac{\omega_o}{\beta} = Q \qquad (1.57)$$

Putting (1.54) and (1.57) together we have

$$Q = 2\pi \frac{\overline{E_{stored}}(\omega_o)}{\overline{E_{loss}}(\omega_o)} = \frac{\omega_o}{BW} \qquad (1.58)$$

The quality factor thus binds together two seemingly different properties of the oscillator, the selectivity of its resonance curve and the rate of energy loss. The higher the Q of the oscillator, the smaller its resonance bandwidth and the more slowly it dissipates energy.

1.8 Linear Models for LC Oscillators

The simplest model of a practical LC oscillator consists of a tank with positive feedback around it [7]. The lossy tank is a damped harmonic oscillator discussed in Sects. 1.4 and 1.5, while the positive feedback effectuates energy restoration to keep the oscillation going. This abstraction treats oscillators as linear positive feedback systems. Equivalently, we may treat oscillators as a negative resistor across a lossy tank, where the negative resistor plays the role of an energy-restoring element. Either way, we start from a damped harmonic oscillator (lossy tank), and by introducing an energy-restoring mechanism, we move toward a simple harmonic oscillator model (lossless tank). Oscillators are inherently nonlinear systems, and linear analysis methods have limited use in predicting the oscillation amplitude. They, however, help predict the oscillation frequency and the conditions for starting the oscillation and allow us to derive an approximate estimate of the oscillator noise due to the loss of the tank.

The feedback model approach is based on expressing the conditions for stability of a feedback system, known as the Barkhausen criteria [6]. These, in simple terms, state that a feedback system oscillates if the loop gain is unity and there is no phase shift around the loop. A general model of an oscillator as a feedback system is shown in Fig. 1.11. The tank voltage is sampled by the linear voltage-controlled current source g_m and is re-injected into the tank as a current. The closed-loop transfer of this unity feedback system is given by

$$\frac{v_{out}}{v_{in}} = \frac{g_m Z(s)}{1 - g_m Z(s)} \qquad (1.59)$$

where $Z(s)$ is the impedance of the damped-LC tank

$$Z(s) = \frac{\frac{s}{C}}{s^2 + \frac{1}{RC}s + \frac{1}{LC}} \qquad (1.60)$$

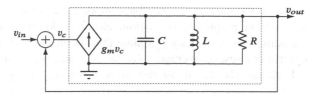

Fig. 1.11 Oscillator feedback model. The oscillator output is sampled and fed back to the input

Fig. 1.12 Locus of the poles of (1.61) as gm is varied from zero to infinity. For $g_m = 1/R$, the two poles are located on the imaginary axis

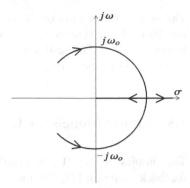

Substituting the expression for the tank impedance into the voltage transfer equation, we obtain

$$\frac{v_{out}}{v_{in}} = \frac{\frac{g_m}{C}s}{s^2 + (\frac{1}{R} - g_m)\frac{1}{C}s + \frac{1}{LC}} \tag{1.61}$$

To find the condition for oscillation, we plot the locus of the poles of the transfer function as gm is increased from zero to infinity in Fig. 1.12. Oscillations occur for $g_m R = 1$ and the angular frequency is $\omega_o = 1/\sqrt{LC}$. The oscillations grow for $g_m R$ slightly higher than one, while the oscillations decay for $g_m R$ slightly lower than one. Oscillators have no input so in Fig. 1.11 v_{in} models noise that helps to startup the oscillations. After starting up, the oscillations are maintained even if v_{in} is removed.

For a reliable startup, g_m has to be larger than $1/R$ [7]. As the oscillation amplitude grows, the signal v_{out} becomes large enough for amplitude limiting due to nonlinearity to kick in. Amplitude limiting regulates the loop gain, effectively ensuring that, on average, $gm = 1/R$, thus maintaining the oscillation and defining the oscillation amplitude. This linear model does not capture the mechanism responsible for amplitude limiting. However, the linear feedback model helps predict the conditions for oscillation, i.e., the minimum value of g_m for the oscillations to start. It also accurately predicts the oscillation frequency but does not account for the fact that the oscillation frequency is amplitude dependent.

An alternative point of view is the negative resistance model shown in Fig. 1.13. The voltage-controlled current source is connected in parallel with the tank in such a way as to present a negative resistance. Oscillations are maintained by making $1/g_m = R$.

As in the case of the feedback model, the negative resistance model fails to account for amplitude regulation. However, knowing that amplitude regulation happens indeed, we may postulate that the mechanism responsible for amplitude control ensures, on average, $1/g_m = R$. Thus, on average, the negative resistance $-1/g_m$ cancels out the loss resistor R as depicted in Fig. 1.14.

Fig. 1.13 Oscillator negative resistance model. The linear negative resistance element compensates for the energy loss in the resistor R

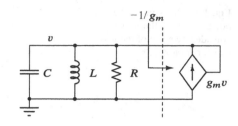

Fig. 1.14 Oscillator negative resistance model. The mechanism responsible for amplitude control ensures that, on average, the negative resistance $-1/g_m$ and the loss resistor R cancel out

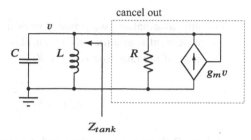

Under the assumption of cancellation, it is interesting to estimate the impedance looking into the tank for frequencies very close to the tank resonance frequency $\omega_o = 1/\sqrt{LC}$ [7] . With reference to Fig. 1.14 we have

$$Y_{tank} = \frac{s^2 LC + 1}{sL} \tag{1.62}$$

Therefore

$$Z_{tank} = \frac{s/C}{s^2 + \omega_o^2} = \frac{j\omega/C}{\omega_o^2 - \omega^2} \tag{1.63}$$

The square magnitude of the tank impedance takes the form

$$|Z_{tank}|^2 = \frac{\omega^2/C^2}{(\omega_o^2 - \omega^2)^2} \tag{1.64}$$

For frequencies very close to the natural frequency ω_o, there is little error introduced by replacing ω by ω_o in the numerator and simplifying the term $(\omega_o^2 - \omega^2)$ in the denominator as $(\omega_o{}^2 - \omega^2) = (\omega_o + \omega)(\omega_o - \omega) \approx 2\omega_o(\omega_o - \omega)$ giving

$$|Z_{tank}|^2 = \frac{1}{(2C\Delta\omega)^2} \tag{1.65}$$

The impedance looking into the tank is inversely proportional to $\Delta\omega$. Equation (1.65) provides an alternative insight into oscillator operation. The positive feedback amplifies internally generated noise resulting in a sharp noise peak at the resonance frequency [8].

Fig. 1.15 To estimate the noise of the oscillator, we consider the noise of the loss resistor R on Z_{tank}

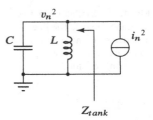

With this information, we proceed to estimate the oscillator noise. The negative resistance element is assumed to be noiseless. We consider the single-sided current noise density of the tank loss resistor R

$$S_{i_n}(f) = \frac{\overline{i_n^2}}{df} = \frac{4kT}{R} \tag{1.66}$$

k and T are Boltzmann's constant and the absolute temperature, respectively. As is depicted in Fig. 1.15, we express the tank noise by the product

$$S_{v_n}(f) = S_{i_n}(f) \cdot |Z_{tank}|^2 \tag{1.67}$$

Equations (1.66) and (1.67) follow from the equipartition and the fluctuation-dissipation theorems and are discussed in Sects. 3.2 and 3.3. Substituting (1.65) and (1.66) into (1.67) gives for the oscillator noise power spectral density

$$S_n(\Delta\omega) = \frac{kT}{R\,C^2(\Delta\omega)^2} \tag{1.68}$$

The power spectral density of the oscillator noise is inversely proportional to the tank resistance and the square of the tank capacitance. Furthermore, it is shaped as $1/\Delta\omega^2$ where $\Delta\omega$ is the frequency offset from the carrier. As is shown in Chap. 3, Eq. (1.68) overestimates noise by a factor of two. It also predicts that the oscillator noise increases without bounds as the frequency offset $\Delta\omega$ goes to zero. This is, of course, not physical and is also revised in Chap. 3.

To obtain a measure of the spectral purity of the oscillator, it is customary to compare the oscillator noise spectral density at a frequency offset $\Delta\omega$ from the carrier to the carrier power $A_1^2/2$

$$\frac{S_n(\Delta\omega)}{A_1^2/2} = \frac{2kT}{R\,C^2 A_1^2(\Delta\omega)^2} \tag{1.69}$$

Equation (1.69) reveals that the oscillator linewidth is inversely proportional to the square of the oscillation amplitude.

The fundamental limitation of linear oscillator models is that they fail to capture the vital characteristic of amplitude regulation that practical oscillators exhibit. Once set in motion, such systems oscillate at their internal rhythm with an amplitude entirely defined by their internal structure. Any small external disturbance temporarily unsettles their motion; they then resettle down gradually to their regular oscillation [4]. On the other hand, the oscillation amplitude of a linear oscillator depends entirely on the initial conditions, and any energy exchange with the environment alters the oscillation amplitude permanently. Furthermore, the abstraction that an energy-restoring mechanism around a lossy tank results in a lossless tank is an approximation. We show in Chap. 3 that the resulting network exhibits finite Q, albeit much higher than the Q of the lossy tank. The interplay and balance between energy dissipation and energy restoration regulates the oscillation amplitude and keeps the oscillation going. To see this, we need to introduce a more suitable model, the self-sustained oscillator.

Chapter 2
Self-Sustained Oscillators

2.1 Introduction

In this chapter, we introduce the more suitable for our purposes model of the self-sustained oscillator. We demonstrate that nonlinearity is essential in obtaining oscillations independent of the initial conditions. We discuss a paradigmatic self-sustained oscillator model, the Van der Pol oscillator, formulate the Van der Pol equation, and present techniques for its solution. Finally, we demonstrate the versatility of the Van der Pol oscillator in modeling realistic oscillators.

2.2 Amplitude Regulation

Despite their simplicity and universal application, the simple harmonic oscillator models presented in the last chapter fail to capture the fundamental characteristic that self-sustained oscillators exhibit, that of amplitude regulation. By experimenting with such systems, we may observe that once set in motion by an external agent, provided an appropriate energy source, and ignoring any synchronization effects, they oscillate at their internal rhythm with an amplitude that depends only on their internal structure. Additionally, any—adequately small—external disturbance unsettles them only temporarily. They then resettle down gradually to their regular oscillation [4]. On the other hand, the oscillation amplitude of the simple harmonic oscillator depends entirely on the initial conditions. Furthermore, any energy exchange with the environment permanently alters the simple harmonic oscillator's amplitude. Effectively, the simple harmonic oscillator has no control over its amplitude.

To extend the simple harmonic oscillator model to account for amplitude regulation, we observe that a damping coefficient $\beta > 0$ results in a stable spiral due to energy dissipated in the resistor across the tank. If we place across the tank

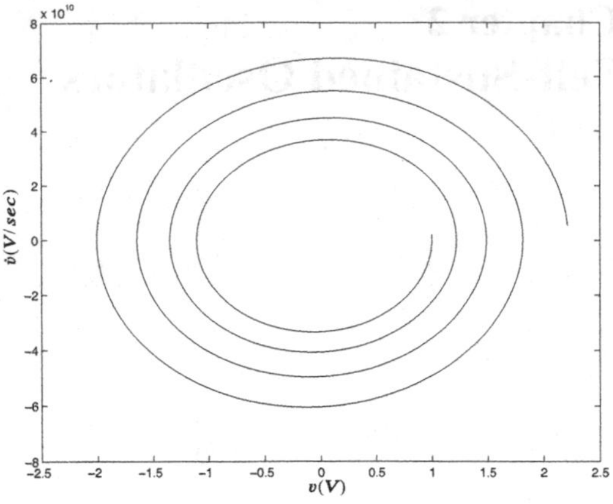

Fig. 2.1 Phase plane path for regenerated LC tank circuit ($C = 1$ pF, $L = 1$ nH, $R = -500\ \Omega$). Energy is added into the tank resulting in an unstable, growing spiral

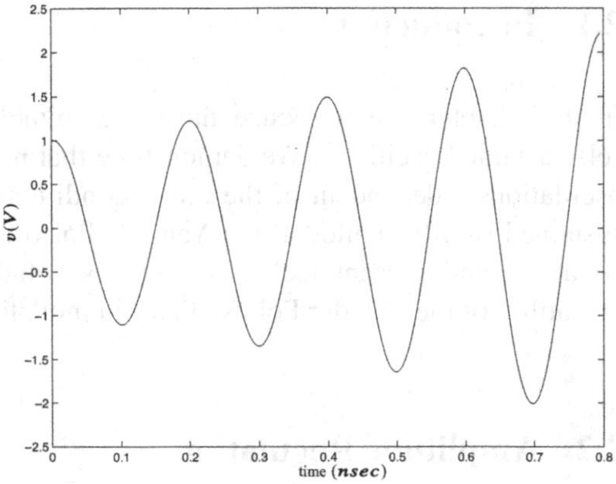

Fig. 2.2 Time response for regenerated LC tank circuit ($C = 1$ pF, $L = 1$ nH, $R = -500\ \Omega$). Energy is added into the talk resulting in an exponentially growing sinewave

a component that delivers energy, we obtain an unstable spiral, since, in this case, $\beta < 0$ as shown in Fig. 2.1. Such a component exhibits negative resistance and, in practice, is implemented with active devices. There is nothing special about negative resistance; it simply designates energy transfer from an external source (power supply) into the tank. As a result, the oscillation amplitude grows with time, as shown in Fig. 2.2.

A self-sustained electric oscillator can thus be modeled by an LC tank with two additional components, as shown in Fig. 2.3. Resistor R models the unavoidable energy loss as in the case of the damped harmonic oscillator. The nonlinear element G adds energy into the tank for small oscillation amplitudes, and the oscillation amplitude grows. On the other hand, for large oscillation amplitudes, energy is dissipated, and the oscillation amplitude decreases. In achieving this kind of behavior, nonlinear elements are required. G is therefore nonlinear as is indicated in Fig. 2.3.

The balance between energy loss and energy injection over the oscillator period regulates the oscillation amplitude, making it independent of any external agent

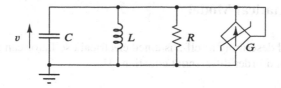

Fig. 2.3 Electrical self-sustained oscillator model. Resistor R dissipates energy, while the nonlinear element represented by G adds energy into the tank for small oscillation amplitudes. For large oscillation amplitudes, energy is dissipated. The balance between energy loss and energy injection over the oscillator period regulates and defines the oscillation amplitude

Fig. 2.4 Self-sustained oscillator limit cycle. Starting the system from any point except the point (0, 0) results in a phase path that gradually converges on the stable limit cycle

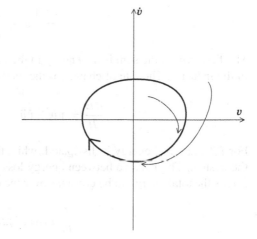

Fig. 2.5 Energy balance results in amplitude regulation in a self-sustained oscillator. The curves represent energy generation and dissipation, plotted against the oscillator amplitude. For low amplitudes, energy generation dominates, resulting in amplitude growth and vice versa

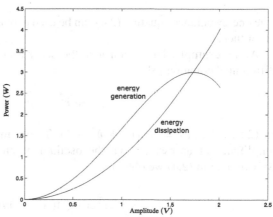

[4]. In the phase plane, the oscillator traverses a closed path known as a stable limit cycle. Any phase path starting inside or outside the limit cycle gradually converges on the limit cycle, as shown schematically in Fig. 2.4. Figure 2.5 depicts the equilibrium between energy loss and regeneration. For small amplitudes, energy regeneration dominates, resulting in amplitude growth and vice versa [4].

2.3 Mathematical Model

A general model describing a self-sustained electrical oscillator can be given in the form of the second-order differential equation [3]

$$\ddot{v} + \beta(v, \dot{v}) + v = 0 \tag{2.1}$$

where $\beta(v, \dot{v})$ is a nonlinear function of v and $\dot{v}$. For simplicity, we have assumed that $\omega_o = 1$. We define the total energy as the sum

$$E = \frac{1}{2}v^2 + \frac{1}{2}\dot{v}^2 \tag{2.2}$$

The first term in the sum is the energy in the capacitor, and the second is the energy in the inductor. The rate of change of the total energy is given by

$$\frac{dE}{dt} = v\dot{v} + \dot{v}\ddot{v} = -\dot{v}\beta(v, \dot{v}) \tag{2.3}$$

For $\dot{v}\beta(v, \dot{v}) > 0$ energy is dissipated, while for $\dot{v}\beta(v, \dot{v}) < 0$ energy is added into the system. The balance between energy loss and energy injection into the system forces the total energy to be constant over the oscillation period T_o

$$\int_{T_o} \dot{v}\beta(v, \dot{v})dt = 0 \tag{2.4}$$

The energy balance equation (2.4) can be used to obtain an estimate of the oscillation amplitude.

As an example, let us consider the self-sustained oscillator described by the differential equation [3]

$$\ddot{v} + (v^2 + \dot{v}^2 - 1)\dot{v} + v = 0 \tag{2.5}$$

In (2.5) $\dot{v}\,\beta(v, \dot{v}) = \dot{v}^2\,(v^2 + \dot{v}^2 - 1)$. To obtain an estimate for the oscillation amplitude, let us assume that the oscillation can be described by $v = A\sin t$. Substituting in (2.4) we obtain

$$\int_T A^2 \cos^2(t)(A^2 - 1)dt = 0 \tag{2.6}$$

which gives the oscillation amplitude $A = 1$.

To find the limit cycle we express (2.5) in phase plane coordinates

$$\dot{x} = y \tag{2.7}$$

$$\dot{y} = (1 - x^2 - y^2)y - x \tag{2.8}$$

where as before $x = v$ and $y = \dot{v}$.

It is furthermore beneficial to express the phase paths in plane polar coordinates r, θ

$$r^2 = x^2 + y^2 \tag{2.9}$$

$$\tan \theta = y/x \tag{2.10}$$

Differentiating (2.9) and (2.10) with respect to time gives

$$r\dot{r} = x\dot{x} + y\dot{y} \tag{2.11}$$

$$\frac{\dot{\theta}}{\cos^2 \theta} = \frac{x\dot{y} - y\dot{x}}{x^2} \tag{2.12}$$

Substituting (2.7), (2.8) into (2.11), (2.12) and using $x = r\cos\theta$ and $y = r\sin\theta$ results in the equations that describe the phase paths in polar coordinates

$$\dot{r} = -r(r^2 - 1)\sin^2\theta \tag{2.13}$$

$$\dot{\theta} = -1 - (r^2 - 1)\sin\theta\cos\theta \tag{2.14}$$

We must have $\dot{r} = 0$ on the limit cycle. By inspection of (2.13) and (2.14) the limit cycle is a circle with radius $r = 1$ traversed clockwise with angular velocity $\dot{\theta} = -1$. The solution of the equation on the limit cycle is $v = \sin(t + \phi)$ where $\phi = constant$ and depends on the initial conditions. As is shown in Figs. 2.6 and 2.7, if we start the system from any point other than the point $(0, 0)$, the system eventually settles on the limit cycle, which in this example is the unit circle. The amplitude is solely determined by the system's internal structure, while the phase depends on the initial conditions. We may thus surmise that self-sustained oscillators control their amplitude but not their phase. This is a general characteristic of such systems [4], which we utilize when analyzing oscillator noise in later chapters.

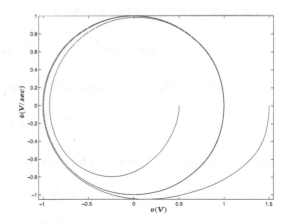

Fig. 2.6 Phase plane paths for self-sustained oscillator example (2.5). Two separate initial points are shown that correspond to $v(0) = 1.5$ and $v(0) = 0.5$, respectively, with $\dot{v}(0) = 0$ for both cases. Both phase paths converge on the unit-circle limit cycle

Fig. 2.7 Time response for
self-sustained oscillator
example (2.5). Two separate
initial points are shown that
correspond to $v(0) = 1.5$ and
$v(0) = 0.5$, respectively, with
$\dot{v}(0) = 0$ for both cases. In
both cases, the system
response converges to a stable
sinusoid with unit amplitude

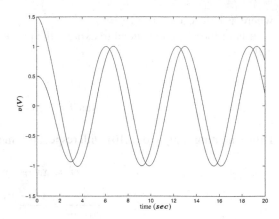

2.4   The Van der Pol Oscillator

The negative resistance is implemented in practice with nonlinear active elements.
Figure 2.8 shows a general model of such an element as a voltage-controlled current
source $G(v)$. The controlling voltage v is the voltage across the tank

$$G(v) = \sum_n g_n v^n \tag{2.15}$$

In the following, we will concentrate on the case where $G(v)$ is odd symmetric.
Furthermore, for simplicity, we disregard powers of v beyond the third. Under these
assumptions $G(v)$ becomes

$$G(v) = g_1 v - g_3 v^3 \tag{2.16}$$

The minus sign in (2.16) results in a compressive characteristic as is required for
amplitude limiting.

The circuit equation for the voltage v in Fig. 2.8 can then be written as

$$g_1 v - g_3 v^3 = \frac{1}{L} \int v \, dt + C \frac{dv}{dt} + \frac{v}{R} \tag{2.17}$$

We subsequently differentiate to remove the integral

$$g_1 \dot{v} - 3 g_3 v^2 \dot{v} = \frac{v}{L} + C \ddot{v} + \frac{\dot{v}}{R} \tag{2.18}$$

where $v = v(t)$ and overdot stands for d/dt. We assume the initial conditions

$$v(0) = A \tag{2.19}$$

$$\frac{dv}{dt}(0) = 0 \tag{2.20}$$

Fig. 2.8 Electronic
self-sustained oscillator
model. The nonlinear element
$G(v)$ is implemented in
practice with active devices

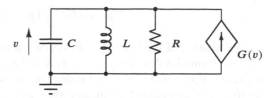

To solve this equation, we first make the following substitutions

$$\omega_1 = \frac{g_1 - \frac{1}{R}}{C} \qquad (2.21)$$

$$\omega_o{}^2 = \frac{1}{LC} \qquad (2.22)$$

$$\rho = \frac{g_3}{g_1 - \frac{1}{R}} \qquad (2.23)$$

Equation (2.18) becomes

$$\ddot{v} - \omega_1(1 - 3\rho v^2)\dot{v} + \omega_o^2 v = 0 \qquad (2.24)$$

To simplify the damping term, we scale the voltage variable by $1/\sqrt{3\rho}$ and express voltage in terms of the dimensionless variable $u = v\sqrt{3\rho}$. Equation (2.24) becomes

$$\ddot{u} - \omega_1(1 - u^2)\dot{u} + \omega_o^2 u = 0 \qquad (2.25)$$

Noting that the friction-less problem has solution $\propto cos(\omega_o t)$, we scale time by $1/\omega_o$, and therefore express time in terms of the dimensionless variable $\tau = \omega_o t$. Using $\dot{u} = \omega_o u'$ and $\ddot{u} = \omega_o^2 u''$ where prime stands for $d/d\tau$ gives

$$u'' - \frac{\omega_1}{\omega_o}(1 - u^2)u' + u = 0 \qquad (2.26)$$

The initial conditions expressed in terms of the new dimensionless variables become $u(0) = \sqrt{3\rho}A$ and $u'(0) = 0$, respectively. The procedure of reformulating the problem in terms of new, dimensionless variables is known as scaling. Scaling is essential in applying perturbation methods that we will use in the following section [10]. Finally, we introduce the parameter

$$\epsilon = \frac{\omega_1}{\omega_o} = \frac{g_1 R - 1}{Q} \qquad (2.27)$$

where $Q = \omega_o RC$ is the tank quality factor. With these substitutions, (2.26) becomes a paradigmatic model in the theory of nonlinear oscillators called the Van der Pol equation [9, 11]

$$u'' + \epsilon(u^2 - 1)u' + u = 0 \qquad (2.28)$$

The solution of the Van der Pol oscillator depends on the value of ϵ. For values of ϵ much smaller than unity ($g_1 \approx 1/R$ and adequately large Q), the Van der Pol oscillator approximates a simple harmonic oscillator. This is because the damping coefficient is sufficiently small, and the Van der Pol equation corresponds to the equation of the simple harmonic oscillator. We, therefore, expect that for small values of ϵ, the oscillation voltage is nearly sinusoidal. For the oscillations to start, we need to ensure that $g_1 > 1/R$ as can be verified by simple linear feedback analysis.

Term ω_1 defined in (2.21) has dimensions of angular frequency and can be regarded as a measure of how fast the oscillator reacts to amplitude perturbations. Any amplitude disturbance dies away exponentially with characteristic time $1/\omega_1$. Since ϵ is proportional to ω_1, small ϵ results in near harmonic output but rather slow amplitude control. In the limiting case of the simple harmonic oscillator, ω_1 equals zero, implying no control over amplitude perturbations. Term ρ defined in (2.23) is proportional to g_3 and inversely proportional to $g_1 - 1/R$. The oscillation amplitude is inversely proportional to the square root of ρ.

Figures 2.9 and 2.10 depict the Van der Pol oscillation behavior for $\epsilon = 0.1$. The amplitude of the oscillator is almost equal to two, and the oscillations are almost simple harmonic [9]. The limit cycle is nearly a circle, but the oscillations take several cycles to start. We may deduce that startup time and amplitude perturbation decay time are related.

As ϵ increases, the limit cycle becomes distinctly non-circular, indicating that the oscillation is no longer a simple harmonic one. This is clearly shown in Figs. 2.11 and 2.12, which correspond to $\epsilon = 1$. The oscillation amplitude remains the same as in the nearly harmonic case, the oscillation period increases somewhat, and startup happens much faster.

Fig. 2.9 Phase plane path for Van der Pol oscillator with $\epsilon = 0.1$. The limit cycle is nearly circular. Initial conditions are $u(0) = 0.5$ and $u'(0) = 0$

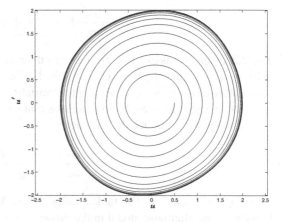

Fig. 2.10 Time response for
Van der Pol oscillator with
$\epsilon = 0.1$. Output is nearly
sinusoidal. Initial conditions
are $u(0) = 0.5$ and $u'(0) = 0$

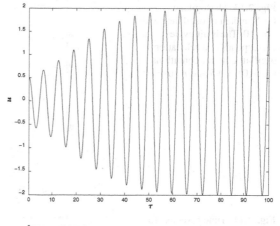

Fig. 2.11 Phase plane path
for Van der Pol oscillator with
$\epsilon = 1$. The limit cycle is
distinctly non-circular. Initial
conditions are $u(0) = 0.5$ and
$u'(0) = 0$

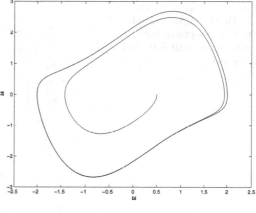

Fig. 2.12 Time response for
Van der Pol oscillator with
$\epsilon = 1$. The oscillation is no
longer a simple harmonic
one. Initial conditions are
$u(0) = 0.5$ and $u'(0) = 0$

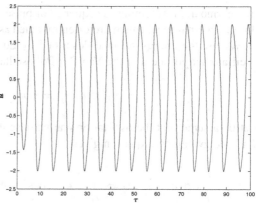

The Van der Pol oscillator becomes a so-called relaxation oscillator for larger
values of ϵ. This is shown in Figs. 2.13 and 2.14 which correspond to $\epsilon = 10$. The
oscillator amplitude remains the same, but the form of the oscillations is nothing

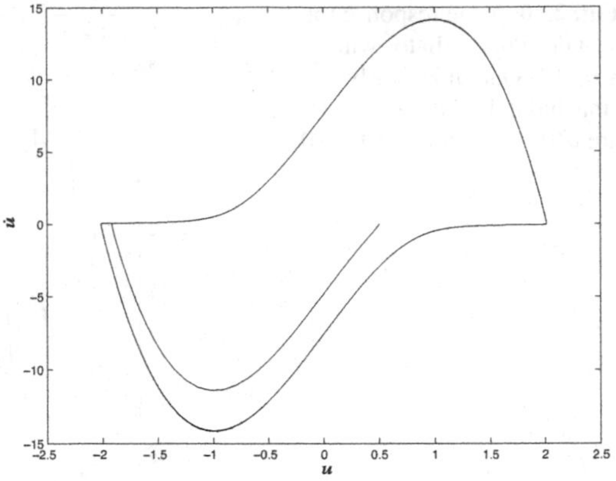

Fig. 2.13 Phase plane path for Van der Pol oscillator with $\epsilon = 10$. The limit cycle corresponds to a relaxation oscillator. Initial conditions are $u(0) = 0.5$ and $u'(0) = 0$

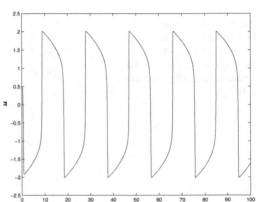

Fig. 2.14 Time response for Van der Pol oscillator with $\epsilon = 10$. The output resembles a sequence of pulses. Initial conditions are $u(0) = 0.5$ and $u'(0) = 0$

like harmonic; it resembles a sequence of pulses. Systems with an "accumulate and dump" internal mechanism are often modeled as relaxation oscillators [9].

We use the energy balance described by Eq. (2.4) to estimate the Van Der Pol oscillator amplitude. For the Van Der Pol oscillator, we have

$$u' \beta(u, u') = \epsilon(u^2 - 1)(u')^2 \tag{2.29}$$

Let us assume that the oscillation can be described approximately by $u = A \sin \tau$. The energy balance equation gives

$$\int_T u' \beta(u, u') = 0 \Rightarrow \int_0^{2\pi} \epsilon A^2 \cos^2\tau \, (A^2 \sin^2\tau - 1)d\tau = 0 \tag{2.30}$$

which yields $A = 2$ in agreement with Figs. 2.10, 2.12, and 2.14.

2.5   Solution by Perturbation

The energy balance equation (2.4) provides information for the oscillation amplitude but does not give us any information about the harmonic components. For this, we need to solve the Van der Pol equation

$$\ddot{u} + \epsilon(u^2 - 1)\dot{u} + u = 0 \tag{2.31}$$

In (2.31) both $u = u(\tau)$ and τ are dimensionless. Overdot stands for $d/d\tau$. We are interested in nearly harmonic oscillators, so we restrict our discussion to the nearly harmonic case ($\epsilon \ll 1$). To solve (2.31), we must either resort to approximation or numerical methods. Foremost among approximation methods are the so-called perturbation methods [10]. Such methods allow us to obtain approximate solutions to problems that have small terms. Furthermore, when the small terms become negligible, the equation becomes that of a well-known problem, which can be solved analytically. In the case of the nearly harmonic Van der Pol equation, ϵ is sufficiently small ($\epsilon \ll 1$). Additionally, for $\epsilon = 0$, (2.31) becomes the equation of the simple harmonic oscillator and is analytically solvable.

Due to the nonlinear character of (2.31), we also need to consider that the oscillation frequency is not constant but depends on the oscillation amplitude. Many nonlinear oscillators exhibit this connection between amplitude and frequency. A typical example from mechanics is the physical pendulum for which it can be proved that for sufficiently small angles of oscillation ϕ, the period of oscillation T_o can be written as

$$T_o = 2\pi \sqrt{\frac{l}{g}} \left(1 + \frac{1}{16}\phi^2 + \cdots \right) \tag{2.32}$$

l is the pendulum length, and g is the acceleration of gravity. The period of oscillation is thus independent of the angular deviation only in the limit of small angles [5].

To capture the dependence of the oscillation frequency on the amplitude, we introduce the distorted time scale $\hat{\tau} = \hat{\omega}\,\tau$ and express the solution of (2.31) in the form of a perturbation series in ϵ [10]

$$u(\hat{\tau}) = u_o(\hat{\tau}) + \epsilon\, u_1(\hat{\tau}) + \epsilon^2 u_2(\hat{\tau}) + \cdots \tag{2.33}$$

The leading term $u_o(\hat{\tau})$ is the solution to the unperturbed problem $u''(\hat{\tau}) + u(\hat{\tau}) = 0$, where prime stands for $d/d\hat{\tau}$. Terms $u_0(\hat{\tau})$, $u_1(\hat{\tau})$, $u_2(\hat{\tau})$, ... are known as perturbation terms. We also express $\hat{\omega}$ as

$$\hat{\omega} = 1 + \epsilon\hat{\omega}_1 + \epsilon^2\hat{\omega}_2 \tag{2.34}$$

where $\hat{\omega}_o$ has been chosen to be unity, the frequency of the solution of the unperturbed problem. We have kept terms up to the second power of ϵ in both perturbation series. Under the scale transformation $\hat{\tau} = \hat{\omega}\,\tau$, Eq. (2.31) becomes

$$u''\hat{\omega}^2 + \epsilon(u^2 - 1)u'\hat{\omega} + u = 0 \qquad (2.35)$$

The procedure we followed to transform (2.31) into (2.35) is known as the Poincaré-Lindstedt method [10, 12].

Substituting the two perturbation series into (2.35) and equating powers of ϵ, we obtain a sequence of linear differential equations for u_o, u_1 and u_2, respectively.

$$u''_o + u_o = 0 \qquad (2.36)$$

$$u''_1 + u_1 = u'_o - u_o^2 u'_o$$

$$-2\hat{\omega}_1 u''_o \qquad (2.37)$$

$$u''_2 + u_2 = u'_1 - u_o^2 u'_1 - 2u_o u_1 u'_o$$

$$+\hat{\omega}_1(u'_o - u_o^2 u'_o - 2u''_1 - \hat{\omega}_1 u''_o)$$

$$-2\hat{\omega}_2 u''_o \qquad (2.38)$$

The initial conditions are $u_o(0) = \sqrt{3\rho}\,A = a_1$ and $u_1(0) = u_2(0) = u'_0(0) = u'_1(0) = u'_2(0) = 0$. The first equation corresponds to the unperturbed problem, while the driving terms in each subsequent equation depend only on the solutions of equations appearing previously in the sequence. The equations can thus be solved in a domino-like fashion.

The first equation is the equation of the simple harmonic oscillator, therefore with initial conditions $u_o(0) = a_o$ and $u_o'(0) = 0$, we obtain

$$u_o = a_o \cos\hat{\tau} \qquad (2.39)$$

Substituting u_o into the equation for u_1 we obtain

$$u''_1 + u_1 = 2\hat{\omega}_1 a_o \cos\hat{\tau} - \left(a_o - \frac{a_o^3}{4}\right)\sin\hat{\tau} + \frac{a_o^3}{4}\sin 3\hat{\tau} \qquad (2.40)$$

Equation (2.40) is the equation of a driven lossless harmonic oscillator. The first two terms on the right side represent harmonic forcing terms at the resonant frequency, resulting in solutions that increase linearly with time while satisfying zero initial conditions. Such diverging terms in the solution are called secular terms and are inconsistent with the physical situation, and the expected solution of (2.31) [10]. To avoid such terms, we set the coefficients of the $\cos\hat{\tau}$ and $\sin\hat{\tau}$ terms equal to zero. Thus

$$\hat{\omega}_1 = 0 \tag{2.41}$$

$$a_o = 2 \tag{2.42}$$

The equation for u_1 then becomes

$$u_1'' + u_1 = 2 \sin 3\hat{\tau} \tag{2.43}$$

which with initial conditions $u_1(0) = 0$ and $u_o'(0) = 0$ gives

$$u_1 = \frac{3}{4} \sin \hat{\tau} - \frac{1}{4} \sin 3\hat{\tau} \tag{2.44}$$

Substituting u_1 into the equation for u_2 we have

$$u_2'' + u_2 = \left(4\hat{\omega}_2 + \frac{1}{4} \right) \cos \hat{\tau} - \frac{3}{2} \cos 3\hat{\tau} + \frac{5}{4} \cos 5\hat{\tau} \tag{2.45}$$

To eliminate the secular term, we equate the coefficient of the $\cos \hat{\tau}$ term to zero and obtain

$$\hat{\omega}_2 = -\frac{1}{16} \tag{2.46}$$

Thus

$$u_2'' + u_2 = -\frac{3}{2} \cos 3\hat{\tau} + \frac{5}{4} \cos 5\hat{\tau} \tag{2.47}$$

which with initial conditions $u_2(0) = 0$ and $u_2'(0) = 0$ gives

$$u_2 = -\frac{13}{96} \cos \hat{\tau} + \frac{3}{16} \cos 3\hat{\tau} - \frac{5}{96} \cos 5\hat{\tau} \tag{2.48}$$

To recapitulate, we have obtained the following perturbation terms

$$u_o = 2 \cos \hat{\tau} \tag{2.49}$$

$$u_1 = \frac{3}{4} \sin \hat{\tau} - \frac{1}{4} \sin 3\hat{\tau} \tag{2.50}$$

$$u_2 = -\frac{13}{96} \cos \hat{\tau} + \frac{3}{16} \cos 3\hat{\tau} - \frac{5}{96} \cos 5\hat{\tau} \tag{2.51}$$

and

$$\hat{\omega}_1 = 0 \tag{2.52}$$

$$\hat{\omega}_2 = -\frac{1}{16} \tag{2.53}$$

Therefore the scaled voltage $u(\hat{t})$ can be expressed as

$$u(\hat{t}) = 2\cos\hat{t} + \epsilon\left[\frac{3}{4}\sin\hat{t} - \frac{1}{4}\sin 3\hat{t}\right]$$

$$+\epsilon^2\left[-\frac{13}{96}\cos\hat{t} + \frac{3}{16}\cos 3\hat{t} - \frac{5}{96}\cos 5\hat{t}\right] \tag{2.54}$$

while the scaled oscillation frequency is

$$\hat{\omega} = 1 - \frac{1}{16}\epsilon^2 \tag{2.55}$$

Referring back to the original voltage and time variables $v(t)$ and t, respectively, we have

$$v(t) = \sqrt{\frac{1}{3\rho}}\left[2\cos\omega t + \epsilon\left(\frac{3}{4}\sin\omega t - \frac{1}{4}\sin 3\omega t\right) + \right.$$

$$\left. \epsilon^2\left(-\frac{13}{96}\cos\omega t + \frac{3}{16}\cos 3\omega t - \frac{5}{96}\cos 5\omega t\right)\right] \tag{2.56}$$

and for the oscillation frequency

$$\omega = \left(1 - \frac{1}{16}\epsilon^2\right)\omega_o \tag{2.57}$$

where

$$\epsilon = \frac{g_1 R - 1}{Q} \tag{2.58}$$

$$\rho = \frac{g_3 R}{g_1 R - 1} \tag{2.59}$$

Assuming that $\epsilon \ll 1$ we may approximate (2.56) as

$$v(t) \approx A_1\cos(\omega t) - A_3\sin(3\omega t) \tag{2.60}$$

where the amplitudes of the first and third harmonics are given by the simplified expressions

$$A_1 = 2\sqrt{\frac{1}{3\rho}} = \sqrt{\frac{4(g_1 R - 1)}{3 g_3 R}} \tag{2.61}$$

$$A_3 = \frac{\epsilon}{4}\sqrt{\frac{1}{3\rho}} = \frac{3 g_3 R}{32 Q} A_1{}^3 \tag{2.62}$$

Figures 2.15, 2.16, and 2.17 compare the predictions of Eqs. (2.61), (2.62), and (2.57) to numerical solution results for $R = 500 \ \Omega$, $C = 1$ pF, $L = 1$ nH, and $g_3 = 1$ mA/V^3 with g_1 varied from 3 mS to 10 mS.

Equation (2.61) reveals that the amplitude of the first harmonic increases by increasing the small-signal transconductance g_1. Increasing g_1 also increases the amplitude of the third harmonic, as shown in Fig. 2.16. The ratio of the third harmonic amplitude to the first harmonic amplitude is given by

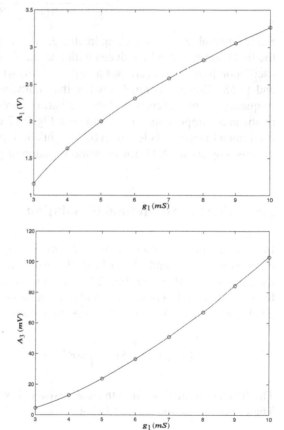

Fig. 2.15 Van der Pol oscillator first harmonic amplitude. Comparison between numerical solution (circles) and Eq. (2.61) (solid line) for $R = 500 \ \Omega$, $C = 1$ pF, $L = 1$ nH, and $g_3' = 1$ mA/V^3 with g_1 varied from 3 mS to 10 mS

Fig. 2.16 Van der Pol oscillator third harmonic amplitude. Comparison between numerical solution (circles) and Eq. (2.62) (solid line) for $R = 500 \ \Omega$, $C = 1$ pF, $L = 1$ nH, and $g_3 = 1$ mA/V^3 with g_1 varied from 3 mS to 10 mS

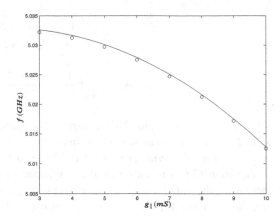

Fig. 2.17 Van der Pol oscillator frequency. Comparison between numerical solution (circles) and Eq. (2.57) (solid line) for $R = 500\ \Omega$, $C = 1$ pF, $L = 1$ nH and $g_3 = 1$ mA/V^3 with g_1 varied from 3 mS to 10 mS

$$\frac{A_3}{A_1} = \frac{g_1 R - 1}{8Q} = \frac{\epsilon}{8} \tag{2.63}$$

Since for reliable startup, we require that $g_1 R > 1$, the amplitude ratio of the third to the first harmonic can be reduced with adequately large tank quality factor Q. The oscillation frequency decreases as g_1 (and thus ϵ) increases as is shown in (2.57) and (2.58). Since ϵ is much smaller than unity, we conclude that the oscillation frequency is a weak function of the oscillation amplitude and is defined to first order by the tank components. This is shown in Fig. 2.17 where the overall variation in the oscillation frequency is less than 0.4%, while in Fig. 2.15, the oscillation amplitude increases by about 180% for the same variation in g_1.

2.6   Alternative Method of Solution

In this section, we present an alternative method to obtain the approximate expressions (2.61) and (2.62) [13]. We work on the Van der Pol self-sustained oscillator model shown in Fig. 2.18. We start the analysis by opening the positive feedback loop at the control terminal of the nonlinear current source G and applying a voltage $A_1 \cos(\omega t)$. The current out of the transconductor becomes

$$i = \left(A_1 g_1 - \frac{3}{4} A_1^{\,3} g_3 \right) \cos(\omega t) - \frac{1}{4} A_1^{\,3} g_3 \cos(3\omega t) \tag{2.64}$$

The first harmonic flows into the tank resistor R, while the third flows into the LC. The impedance of the parallel LC at 3ω is

$$Z_{LC}|_{3\omega} = -j\frac{3}{8\omega C} \tag{2.65}$$

Fig. 2.18 Electrical
self-sustained oscillator
model

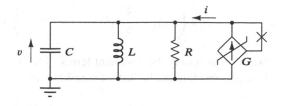

This gives the tank voltage

$$v = \left(A_1 g_1 - \frac{3}{4} A_1^3 g_3\right) R \cos(\omega t) - \frac{3}{32\omega C} A_1^3 g_3 \sin(3\omega t) \qquad (2.66)$$

Equation (2.66) can be further written as

$$v = \left(A_1 g_1 - \frac{3}{4} A_1^3 g_3\right) R \cos(\omega t) - \frac{3}{32 Q} A_1^3 g_3 R \sin(3\omega t) \qquad (2.67)$$

where we have used the tank quality factor expression $Q = \omega RC$.
   The first harmonic amplitude A_1 can be found by equating

$$A_1 = \left(A_1 g_1 - \frac{3}{4} A_1^3 g_3\right) R \qquad (2.68)$$

which gives

$$A_1 = \sqrt{\frac{4(g_1 R - 1)}{3 g_3 R}} \qquad (2.69)$$

The oscillator carrier therefore becomes

$$v \approx A_1 \cos(\omega t) - A_3 \sin(3\omega t) \qquad (2.70)$$

The amplitude of the 3rd harmonic can be obtained from (2.67) and (2.69) giving

$$A_3 = \frac{(g_1 R - 1)}{8 Q} A_1 \qquad (2.71)$$

These results are the same as the simplified expressions (2.60), (2.61), and (2.62).
This method does not predict the slight reduction in the oscillation frequency
with amplitude given in (2.57). It can nevertheless be considered a reasonable
approximation of the Van der Pol oscillator solution for $\epsilon \ll 1$.
   The transconductor current can be obtained by substituting (2.70) into $i = g_1 v - g_3 v^3$. This gives

$$i \approx \left(A_1 g_1 - \frac{3}{4} A_1{}^3 g_3 \right) \cos(\omega t) - \frac{1}{4} A_1{}^3 g_3 \cos(3\omega t) \tag{2.72}$$

We have kept only the dominant terms of the first and third harmonic components. Using (2.69) this can be further simplified as

$$i \approx \frac{A_1}{R} \cos(\omega t) - \frac{A_1}{R} \frac{(g_1 R - 1)}{3} \cos(3\omega t) \tag{2.73}$$

The first and third harmonic transconductor current amplitudes are thus given by

$$I_1 = \frac{A_1}{R} \tag{2.74}$$

$$I_3 = \frac{A_1}{R} \cdot \frac{g_1 R - 1}{3} \tag{2.75}$$

The ratio of the third harmonic current amplitude to the first harmonic current amplitude is given by

$$\frac{I_3}{I_1} = \frac{g_1 R - 1}{3} \tag{2.76}$$

2.7 NMOS-Only Oscillator

The Van der Pol oscillator captures the vital characteristics of self-sustained oscillators. At the same time, using it to model realistic oscillators is relatively straightforward.

We consider the NMOS-only oscillator shown in Fig. 2.19 to demonstrate this. The tank components are chosen to center the oscillator carrier at 5 GHz. The 250 Ω resistor models tank loss. For the MOS devices, we utilize a VerilogA MOS model that represents the drain current using a single expression that is continuous and valid in all regions of operation [14]

$$I_D = 2 \frac{W}{L} K_p n \phi_t{}^2 \cdot$$
$$\left[ln^2 \left(1 + e^{\frac{V_{GS} - V_T}{2n\phi_t}} \right) - ln^2 \left(1 + e^{\frac{V_{GS} - V_T - n V_{DS}}{2n\phi_t}} \right) \right] \tag{2.77}$$

W, L, and K_p are the device width, length, and transconductance parameter, respectively. V_{GS}, V_{DS} are the Gate–Source and Drain–Source voltages. ϕ_t is the thermal voltage, V_T is the threshold voltage, and n is the subthreshold slope.

The individual device currents I_1 and I_2 can be expressed in terms of the differential and common-mode currents I_d and I_{cm} as

Fig. 2.19 Realistic oscillator with NMOS transconductor. The tank components center the oscillator carrier at 5 GHz. The 250 Ω resistor models tank loss. NMOS devices are modeled according to (2.77)

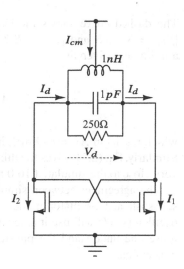

Fig. 2.20 Differential current I_d against differential voltage V_d for the oscillator shown in Fig. 2.19. Continuous curve: Simulation. Dashed curve: Estimation of differential current using Eq. (2.79) with $|g_1| = 8.7$ mS and $|g_3| = 8.7$ mA/V^3. For the devices we have used $W = 50\ \mu$m, $L = 100$ nm, $K_p = 200\ \mu$A/V^2, $V_T = 0.35$ V, and $n = 1$. Supply voltage is 0.5 V

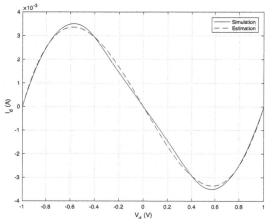

$$I_{1,2} = \frac{I_{cm}}{2} \pm I_d \tag{2.78}$$

In Fig. 2.19, differential current I_d flows into the tank differential impedance, while the common-mode current I_{cm} flows from the supply via the tank common-mode impedance into the transconductor and terminates to ground. We model the differential current according to (2.16) as

$$I_d = g_1 V_d + g_3 V_d{}^3 \tag{2.79}$$

where V_d is the differential voltage. g_1 and g_3 have opposite signs as the transconductor exhibits limiting behavior. In Fig. 2.20, the continuous curve shows the simulated differential transconductor current I_d against differential voltage V_d.

The dashed curve shows the expected differential current using Eq. (2.79) with $|g_1| = 8.7$ mS and $|g_3| = 8.7$ mA/V^3. The expected oscillator's first harmonic amplitude is given by

$$A_1 = \sqrt{\frac{4(g_1 R - 1)}{3 g_3 R}} \tag{2.80}$$

which results in $A_1 = 849$ mV, in close agreement to the simulated 851 mV value. Similarly, Eq. (2.71) gives the third harmonic amplitude, resulting in $A_3 = 15.8$ mV, very close to the simulated 16.0 mV value.

The agreement between simulation and analysis indicates that the Van der Pol self-sustained oscillator model depicted in Fig. 2.18 accurately models realistic oscillators. We will use it extensively in the following chapters to model oscillator noise and understand the physical mechanisms that govern noise to phase noise conversion.

Chapter 3
Noise in LC Oscillators

3.1 Introduction

The spectrum of a noiseless oscillator has the form of a delta function. All energy is concentrated at the oscillation frequency. Real oscillators, however, exhibit a broadening of their spectral line due to noise. A noisy carrier of amplitude A_1 and angular frequency ω_o can be expressed as

$$v(t) = A_1(1 + v_t)\sin(\omega_o t + \theta_t) \tag{3.1}$$

where v_t and θ_t are functions of time and represent amplitude and phase noise, respectively. The amplitude-limiting behavior present in any self-sustained oscillator suppresses amplitude fluctuations. Phase fluctuations thus dominate oscillator noise. This chapter concentrates on oscillator noise and investigates the fundamental mechanisms behind lineshape broadening.

3.2 Equipartition and Fluctuation-Dissipation Theorems

In Sect. 1.4, we considered an isolated damped harmonic oscillator. The discussion suggested that due to energy dissipation on the tank loss resistor, the oscillator energy decreases exponentially with time and eventually decays to zero. In reality, an oscillator is never isolated from its environment. It continuously interacts with it, eventually reaching a state of equilibrium. In thermal equilibrium, the oscillator and its surroundings are characterized by absolute temperature T as is shown in Fig. 3.1 [15].

Any thermodynamic quantity in thermal equilibrium exhibits thermal fluctuations around its average value. In Fig. 3.1, these thermal fluctuations can be associated with the thermal motion of the electrons in the resistor R at absolute

Fig. 3.1 Damped harmonic
oscillator in thermal
equilibrium with its
environment at temperature T

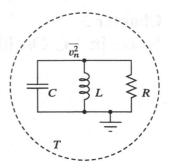

temperature T, resulting in a fluctuating current through the LC tank. The thermal
fluctuations average zero, but their mean square value is not zero. A random voltage
$\overline{v_n^2}$ thus appears across the tank.

The equipartition theorem predicts that the average energy associated with
fluctuations per degree of freedom in thermal equilibrium at temperature T is $kT/2$
where k is Boltzmann's constant. The oscillator has two degrees of freedom. When
its energy decays and becomes comparable to kT, it stops decreasing, and due to
the interaction with the environment, it fluctuates around this average value. Two
requirements must be fulfilled for the equipartition theorem to apply [15]. The first
is that the system under observation obeys the Boltzmann distribution, that is, the
probability that the system is at a state with energy E is given by

$$P(E) = \frac{1}{Z} e^{-\frac{E}{kT}} \tag{3.2}$$

Z is the partition function whose value is calculated so that the sum of the
probabilities over all the allowable energy states is unity (normalization condition)

$$\int_{-\infty}^{+\infty} P(E) dE = 1 \Rightarrow Z = \int_{-\infty}^{+\infty} e^{-\frac{E}{kT}} dE \tag{3.3}$$

The second requirement is that the energy per degree of freedom x obeys a quadratic
dependence $E(x) = \alpha x^2$ where α is a constant. Then the average energy per degree
of freedom is given by

$$\overline{E} = \frac{1}{Z} \int_{-\infty}^{+\infty} \alpha x^2 e^{-\frac{\alpha x^2}{kT}} dx \tag{3.4}$$

Equations (3.3) and (3.4) give

$$\overline{E} = \frac{\int_{-\infty}^{+\infty} \alpha x^2 e^{-\frac{\alpha x^2}{kT}} dx}{\int_{-\infty}^{+\infty} e^{-\frac{\alpha x^2}{kT}} dx} = \frac{kT}{2} \tag{3.5}$$

Fig. 3.2 Spontaneous fluctuations are modeled with the noise current $\overline{i_n^2}$ across the tank. The resistor R in parallel with the LC tank in the lower part of the Figure is noiseless

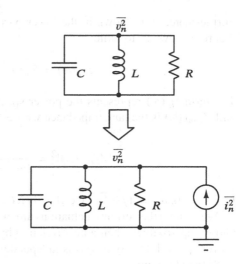

The fluctuation-dissipation theorem asserts a connection between thermal fluctuations and dissipation, i.e., energy-sapping due to frictional phenomena [16]. The ability of a system to absorb and dissipate energy is proportional to the magnitude of thermal fluctuations in thermal equilibrium. The response of a system to an external disturbance and the internal fluctuation of the system in the absence of the disturbance are therefore linked. An impedance or an admittance function may characterize such a response (see (3.8) in the following section). The theorem allows us to describe the internal fluctuations of the damped harmonic oscillator shown in the upper part of Fig. 3.2 by its response to an external noise source as is depicted in the lower part of the figure. The resistor R, in parallel with the LC tank in the lower part of Fig. 3.2, is noiseless. The equivalent network shown in the lower part of Fig. 3.2 is a noise-driven damped harmonic oscillator.

3.3 Damped Harmonic Oscillator Noise

Since the energy stored in the tank capacitor in Fig. 3.2 depends on the square of the voltage across it, the equipartition theorem yields

$$\overline{E} = \frac{1}{2} C \overline{v_n^2} = \frac{kT}{2} \Rightarrow \overline{v_n^2} = \frac{kT}{C} \tag{3.6}$$

We write the power spectral density of the voltage fluctuations as $S_{v_n}(\omega)$. Integrating $S_{v_n}(\omega)$ across frequency gives the total noise power kT/C (Parseval's theorem)

$$\frac{1}{2\pi} \int_{-\infty}^{\infty} S_{v_n}(\omega) d\omega = \frac{kT}{C} \tag{3.7}$$

Furthermore, as is shown in the lower part of Fig. 3.2, the fluctuation-dissipation
theorem allows us to write

$$S_{v_n}(\omega) = S_{i_n}(\omega) \cdot |Z_{tank}(\omega)|^2 \tag{3.8}$$

The term $S_{i_n}(\omega)$ represents the power spectral density of the current fluctuations,
and $Z_{tank}(\omega)$ is the tank impedance whose square magnitude is given by

$$|Z_{tank}(\omega)|^2 = \frac{(\omega/C)^2}{(\omega_o{}^2 - \omega^2)^2 + (\beta\omega)^2} \tag{3.9}$$

In Eq. (3.9), $\omega_o = 1/\sqrt{LC}$ and $\beta = 1/RC$.

We assume the current fluctuations are much faster than any frequency of interest,
and we approximate their autocorrelation by a delta function. Therefore, the current
power spectral density $S_{i_n}(\omega)$ is independent of the frequency. Equations (3.7) and
(3.8) therefore give

$$S_{i_n}(\omega) \cdot \frac{1}{2\pi} \int_{-\infty}^{\infty} |Z_{tank}(\omega)|^2 d\omega = \frac{kT}{C} \tag{3.10}$$

Finally, taking into account that

$$\int_{-\infty}^{\infty} \frac{\omega^2 d\omega}{(\omega_o^2 - \omega^2)^2 + (\beta\omega)^2} = \frac{\pi}{\beta} \tag{3.11}$$

we obtain

$$S_{i_n}(f) = \frac{\overline{i_n^2}}{df} = \frac{2kT}{R} \tag{3.12}$$

Equation (3.12) is known as Nyquist's noise theorem [17]. The current power
spectral density units are (A^2/Hz). We can thus calculate the power spectral density
of the voltage fluctuations across the tank by substituting Eqs. (3.12) and (3.9) into
(3.8) as

$$S_{v_n}(\omega) = \frac{2kT}{R} \cdot \frac{(\omega/C)^2}{(\omega_o{}^2 - \omega^2)^2 + (\beta\omega)^2} \tag{3.13}$$

The single-sided spectrum is plotted in Fig. 3.3.

Fig. 3.3 Single-sided power spectral density of voltage fluctuation across the tank in Fig. 3.2 with $C = 1$ pF, $L = 1$ nH and $R = 500\ \Omega$

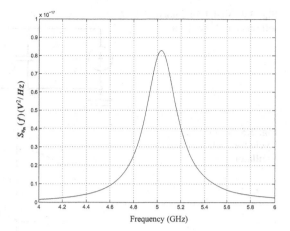

3.4   Self-Sustained Oscillator Phase Noise—Dissipation

We now focus on the self-sustained oscillator depicted in Fig. 3.4. The LC tank defines the oscillation frequency $\omega_o = 1/\sqrt{LC}$ while resistor R accounts for losses. All details of the $i(v)$ nonlinearity are discarded, and it is assumed that due to the energy balance between $i(v)$ and the resistor R, the oscillator amplitude A_1 is stabilized and well-defined as described in Chap. 2.

Over an oscillation cycle, the energy lost due to the deterministic motion of the system is compensated by the energy injected into the system by the nonlinear current source $i(v)$. The noise current source $\overline{i_n^2}$ results in random voltage fluctuations across the tank. In this section, we establish that a self-sustained oscillator can be viewed as a noise-driven damped harmonic oscillator, and we obtain a value for the damping rate and the quality factor.

Noise due to the loss resistor R is modeled by the stationary noise source with noise current density $\overline{i_n^2}/df = 2kT/R$ per the Nyquist noise theorem. The mean square noise voltage across the tank $\overline{v_n^2}$ can be obtained from the equipartition theorem as

$$\overline{E} = \frac{1}{2}C\overline{v_n^2} = \frac{kT}{2} \Rightarrow \overline{v_n^2} = \frac{kT}{C} \tag{3.14}$$

Over time Δt, energy ΔE is dissipated on the resistor R. Therefore

$$\left|\frac{\Delta E}{\Delta t}\right| = \frac{\overline{v_n^2}}{R} = \frac{kT}{RC} \tag{3.15}$$

The quality factor of the oscillator is defined in Chap. 1 as the average energy stored in the tank divided by the average energy loss per cycle per radian

Fig. 3.4 Model for a noisy
self-sustained oscillator

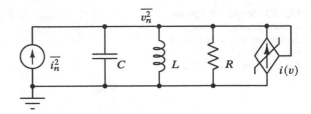

Fig. 3.5 Idealized Limit
Cycle for a self-sustained
oscillator

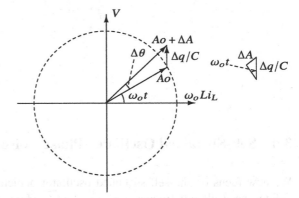

$$Q = 2\pi \frac{\overline{E_{tank}}}{\overline{\Delta E}} = \omega_o \frac{\frac{1}{2}CA_1{}^2}{\frac{1}{2}\left|\frac{\Delta E}{\Delta t}\right|} \tag{3.16}$$

On the right side of Eq. (3.16), we state that only half of the dissipated power $|\Delta E/\Delta t|$ broadens the spectral line of the oscillator. The reasoning behind this can be seen with the aid of the idealized limit cycle in Fig. 3.5 as follows. Assuming that over time Δt, noise charge Δq is injected into the tank, the induced voltage change $\Delta q/C$ in the tank capacitor results in phase and amplitude deviations given by

$$A_1\Delta\theta \approx \frac{\Delta q}{C}\cos(\omega_o t) \tag{3.17}$$

and

$$\Delta A \approx \frac{\Delta q}{C}\sin(\omega_o t) \tag{3.18}$$

respectively. The oscillator carrier is $A_1\sin(\omega_o t)$. The cosine term in (3.17) shows that the phase is most sensitive to noise fluctuations when the carrier crosses zero, while the sine term in (3.18) shows that the amplitude is most sensitive to noise fluctuations when the carrier is at its peak values. We further assume that the amplitude deviations are suppressed by the nonlinear, amplitude-limiting nature of the oscillator infinitely fast. Thus, only the phase deviations result in broadening the

Fig. 3.6 Self-sustained oscillator as a noise-driven lightly damped oscillator. R_θ is given by (3.22). The noise current density $\overline{i_n^2}/df$ is given by $2kT/R$, where R is the tank resistor

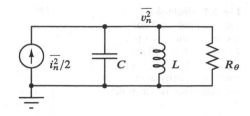

oscillator spectral line. The mean square value of the phase deviations is obtained by squaring and averaging (3.17), giving

$$\overline{\Delta\theta^2} \approx \frac{\overline{\Delta q^2}}{C^2 A_1{}^2}\overline{\cos^2(\omega_o t)} \tag{3.19}$$

Averaging over one period of the oscillation results in

$$\overline{\Delta\theta^2} \approx \frac{\overline{\Delta q^2}/2}{C^2 A_1{}^2} \tag{3.20}$$

which shows that only half of the injected noise charge results in phase fluctuations and, thus, in the broadening of the oscillator linewidth.

Substituting Eq. (3.15) into (3.16) gives

$$Q = Q_t \frac{CA_1{}^2}{kT} \tag{3.21}$$

where $Q_t = \omega_o RC$ is the tank quality factor. Equation (3.21) reveals that the quality factor obtained by placing the nonlinear active element $i(v)$ across the tank is not infinite. It is given by the tank quality factor Q_t enhanced by the signal-to-noise ratio. It follows that the parallel combination of R and $i(v)$ is also not infinite and is given by

$$R_\theta = R\frac{CA_1{}^2}{kT} \tag{3.22}$$

as is shown in Fig. 3.6. We can thus model a self-sustained oscillator as a noise-driven lightly damped oscillator with a loss resistor given by Eq. (3.22). The noise current density $\overline{i_n^2}/df$ in Fig. 3.6 is given by $2kT/R$, where R is the tank resistor.

The spectrum of the self-sustained oscillator in Fig. 3.6 can be estimated by

$$S_\theta(\omega) = \frac{1}{2} \cdot \frac{2kT}{R} \cdot \frac{(\omega/C)^2}{(\omega_o{}^2 - \omega^2)^2 + (\beta_\theta\,\omega)^2} \tag{3.23}$$

In (3.23) ω spans the whole frequency range $(-\infty, \infty)$. The damping rate is given by

Fig. 3.7 Single-sided
self-sustained oscillator
spectrum with $C = 1$ pF,
$L = 1$ nH, $R = 500\ \Omega$,
$A_1 = 1$ V. The x-axis is the
frequency relative to the
oscillator's natural frequency

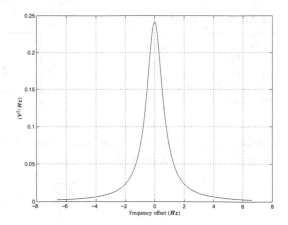

$$\beta_\theta = \frac{1}{R_\theta C} = \frac{1}{RC} \frac{kT}{CA_1^2} \tag{3.24}$$

The single-sided spectrum is plotted in Fig. 3.7. The self-sustained oscillator linewidth is orders of magnitude narrower than the linewidth of the damped harmonic oscillator depicted in Fig. 3.3. Their ratio is given by

$$\frac{R_\theta}{R} = \frac{CA_1^2}{kT} \tag{3.25}$$

Integrating the noise power in (3.23) over the entire frequency range with the help of (3.11) results in

$$\frac{1}{2} \cdot \frac{2kT}{R} \cdot \frac{1}{C^2} \cdot \left(\frac{1}{2\pi} \cdot \frac{\pi}{\beta_\theta}\right) = \frac{A_1^2}{2} \tag{3.26}$$

which is the carrier power. The factor $1/2\pi$ is included since $\omega = 2\pi f$. From (3.26) we obtain

$$\beta_\theta A_1^2 = \frac{kT}{RC^2} \tag{3.27}$$

For frequencies close enough to ω_o, we can simplify (3.23) by replacing ω by ω_o everywhere except in the term $\omega_o^2 - \omega^2$, which is written as $2\omega_o(\omega_o - \omega)$. Therefore

$$S_\theta(\omega) \approx \frac{kT}{4RC^2} \cdot \left[\frac{1}{(\omega_o + \omega)^2 + (\beta_\theta/2)^2} + \frac{1}{(\omega_o - \omega)^2 + (\beta_\theta/2)^2}\right] \tag{3.28}$$

The first term in the sum corresponds to negative frequencies, while the second term corresponds to the positive part of the spectrum. Substituting (3.27) into (3.28) gives

$$S_\theta(\omega) \approx \frac{A_1{}^2 \beta_\theta}{4} \cdot \left[\frac{1}{(\omega_o + \omega)^2 + (\beta_\theta/2)^2} + \frac{1}{(\omega_o - \omega)^2 + (\beta_\theta/2)^2} \right] \qquad (3.29)$$

To obtain a metric of the spectral purity of the oscillator, it is customary to compare the phase noise spectral density (per Hz) at a frequency offset $\Delta\omega = \omega_o - \omega$ from the carrier to the carrier power. We express this ratio as PN, which is shorthand for Phase Noise, by

$$PN = \frac{S_\theta(\omega)}{A_1{}^2/4} \approx \frac{\beta_\theta}{(\Delta\omega)^2 + (\beta_\theta/2)^2} \qquad (3.30)$$

Only positive frequencies are considered in (3.30). In the limit $\Delta\omega \to 0$, Equation (3.30) gives $4/\beta_\theta$ and does not diverge as (1.69).

For frequency offsets much larger than $\beta_\theta/2$, Equation (3.30) can be simplified to the approximation known as Leeson's formula [18]

$$PN \approx \frac{\beta_\theta}{(\Delta\omega)^2} \qquad (3.31)$$

Substituting β_θ from (3.24) gives

$$PN \approx \frac{kT}{RC^2 A_1{}^2 (\Delta\omega)^2} \qquad (3.32)$$

The factor two that appears in (1.69) is not present in (3.32). (3.32) shows that PN is inversely proportional to the tank loss resistor R, and the squares of the oscillation amplitude $A_1{}^2$ and the tank capacitance C^2. Smaller tank loss resistor R results in lower thermal noise injected across the tank. Larger tank capacitance C results in smaller voltage deviation for a given noisy charge injection. Larger carrier amplitude A_1 results in increased carrier power. The behavior as $1/(\Delta\omega)^2$ results in a 6dB reduction in the PN with every doubling of the offset frequency.

3.5   Self-Sustained Oscillator Phase Noise—Fluctuation

In this section, we derive the spectrum of the self-sustained oscillator given in Eq. (3.29)) by considering the phase fluctuations. To do this, we model the time evolution of the random phase θ_t in (3.1) as a one-dimensional random-walk process [16, 17].

In the following, we drop the subscript t and denote the random phase as θ. We assume the random phase steps have equal magnitude $|\Delta\theta|$ and are equally likely to advance or retard the total phase. Furthermore, we assume that they are uncorrelated from each other. After N such steps, the random component of the total phase becomes

$$\theta = \Delta\theta_1 + \Delta\theta_2 + \ldots + \Delta\theta_N \tag{3.33}$$

θ is normally distributed as it is the outcome of many uncorrelated random variables. In any particular observation of N random phase steps, the resulting overall phase deviation may be either positive or negative. For a large number of similar experiments, it is likely that for every observed positive value of θ, we will also find a corresponding negative value. It is thus reasonable to expect that the average phase is zero

$$\overline{\theta} = 0 \tag{3.34}$$

Equation (3.34) gives us the "expectation" value of θ but does not tell us how far the phase might depart from zero in the positive or the negative direction in each experiment [17]. To obtain this information, we need to consider the variance $\sigma_\theta{}^2$, which, since $\overline{\theta} = 0$, is given by

$$\sigma_\theta{}^2 = \overline{\theta^2} = \overline{(\Delta\theta_1 + \Delta\theta_2 + \ldots + \Delta\theta_N)(\Delta\theta_1 + \Delta\theta_2 + \ldots + \Delta\theta_N)} \tag{3.35}$$

Taking into account that the phase steps are equal in magnitude and uncorrelated, (3.35) gives

$$\sigma_\theta{}^2 = N\,\overline{\Delta\theta^2} \tag{3.36}$$

Assuming further that each phase jump occurs during time Δt and the total observation time is t, the above equation becomes

$$\sigma_\theta{}^2 = \frac{t}{\Delta t}\,\overline{\Delta\theta^2} \tag{3.37}$$

The variance of the phase is proportional to time, which is a characteristic of diffusion processes [16]. The connection between random walk and diffusion is established by the fact that at every step, there is an equal probability that the phase will either increase or decrease. Thus the probability $P_{t+\Delta t}(\theta)$ that the value of the random phase at time $t + \Delta t$ is θ can be expressed as

$$P_{t+\Delta t}(\theta) = \frac{1}{2}P_t(\theta + \Delta\theta) + \frac{1}{2}P_t(\theta - \Delta\theta) \tag{3.38}$$

subtracting $P_t(\theta)$ from both sides gives

$$P_{t+\Delta t}(\theta) - P_t(\theta) = \frac{1}{2}[P_t(\theta + \Delta\theta) - P_t(\theta)] - \frac{1}{2}[P_t(\theta) - P_t(\theta - \Delta\theta)] \tag{3.39}$$

and finally multiplying by $(\Delta\theta)^2$ and dividing by Δt gives the diffusion Equation [16]

$$\frac{\partial P}{\partial t} = \frac{(\Delta\theta)^2}{2\Delta t} \cdot \frac{\partial^2 P}{\partial\theta^2} = D\frac{\partial^2 P}{\partial\theta^2} \tag{3.40}$$

where D is the phase diffusion constant. From Eqs. (3.40) and (3.37) the phase diffusion constant can be written as

$$D = \frac{1}{2}\frac{\overline{\Delta\theta^2}}{\Delta t} = \frac{1}{2}\frac{\sigma_\theta{}^2}{t} \tag{3.41}$$

The diffusion equation describes physical phenomena where a quantity moves from regions of high concentration toward areas of lower concentration [19]. The flow at every point is proportional to the diffusion constant and the concentration slope. The total amount of quantity is conserved. Furthermore, given enough time, the concentration settles down to a steady-state, independent of time. In the case of the oscillator phase, as the oscillator exerts no control on it, Eq. (3.40) quantifies its diffusion and spreading out from an initial delta distribution.

The mean square value of the phase deviations is given by (3.20)

$$\overline{\Delta\theta^2} \approx \frac{\overline{\Delta q^2}/2}{C^2 A_1{}^2} \tag{3.42}$$

Using Nyquist's noise theorem, we can express the mean square value of the charge deviations as

$$\overline{\Delta q^2} = \frac{\overline{i_n^2}}{\Delta f}\Delta t = \frac{2kT}{R}\Delta t \tag{3.43}$$

Combining (3.42) and (3.43) gives

$$\overline{\Delta\theta^2} \approx \frac{kT}{RC^2 A_1{}^2}\Delta t \tag{3.44}$$

Substituting Eq. (3.44) into (3.41) gives for the phase diffusion constant

$$D = \frac{1}{2}\frac{kT}{RC^2 A_1{}^2} \tag{3.45}$$

The phase damping rate is given by (3.24)

$$\beta_\theta = \frac{kT}{RC^2 A_1{}^2} \tag{3.46}$$

Therefore

$$D = \frac{\beta_\theta}{2} \tag{3.47}$$

showing that the diffusion constant equals one-half of the phase damping rate.

To obtain the oscillator spectrum, we write the oscillator carrier and calculate its autocorrelation. The power spectrum and the autocorrelation function are linked by the Wiener-Khintchine theorem. They are Fourier transform related [17]. We write the oscillator carrier as

$$V = A_1 \sin(\omega_o t + \theta_t) \tag{3.48}$$

Subscript t denotes time dependency. We have disregarded amplitude noise in (3.48). The autocorrelation is calculated by

$$R_{VV}(\tau) = A_1{}^2 \cdot \overline{\sin(\omega_o t + \theta_t) \sin(\omega_o t + \omega_o \tau + \theta_{t+\tau})} \tag{3.49}$$

which is expanded as

$$R_{VV}(\tau) = \frac{A_1{}^2}{2} \cdot \overline{\cos(\omega_o \tau + \theta_{t+\tau} - \theta_t) - \cos(2\omega_o t + \omega_o \tau + \theta_{t+\tau} + \theta_t)} \tag{3.50}$$

The second term in (3.50) is a cosine with frequency $2\omega_o$, so its average value is zero. Therefore (3.50) is written as

$$R_{VV}(\tau) = \frac{A_1{}^2}{2} \cdot \overline{\cos(\omega_o \tau) \cos(\theta_{t+\tau} - \theta_t) - \sin(\omega_o \tau) \sin(\theta_{t+\tau} - \theta_t)} \tag{3.51}$$

where we have expanded the non-zero cosine term. Taking the constant terms $\cos(\omega_o \tau)$ and $\sin(\omega_o \tau)$ out gives

$$R_{VV}(\tau) = \frac{A_1{}^2}{2} \cos(\omega_o \tau) \cdot \overline{\cos(\theta_{t+\tau} - \theta_t)} - \frac{A_1{}^2}{2} \sin(\omega_o \tau) \cdot \overline{\sin(\theta_{t+\tau} - \theta_t)} \tag{3.52}$$

The second term in (3.52) is zero as the term $\sin(\theta_{t+\tau} - \theta_t)$ is an odd function. Therefore

$$R_{VV}(\tau) = \frac{A_1{}^2}{2} \cos(\omega_o \tau) \cdot \overline{\cos(\theta_{t+\tau} - \theta_t)} \tag{3.53}$$

For normally distributed θ we have

$$\overline{\cos \theta} = \int_{-\infty}^{+\infty} \cos \theta \, \frac{1}{\sqrt{2\pi \sigma_\theta{}^2}} e^{\frac{-\theta^2}{2\sigma_\theta{}^2}} \, d\theta = e^{\frac{-\sigma_\theta{}^2}{2}} \tag{3.54}$$

Therefore

$$R_{VV}(\tau) = \frac{A_1{}^2}{2} e^{\frac{-\sigma_\theta{}^2}{2}} \cos(\omega_o \tau) \tag{3.55}$$

Finally, substituting $\sigma_\theta^2 = 2Dt$ from (3.41) results in

$$R_{VV}(\tau) = \frac{A_1^2}{2} e^{-D|\tau|} \cos(\omega_o \tau) \qquad (3.56)$$

where we have introduced the absolute value since autocorrelation is an even function. The autocorrelation function of the carrier reveals information about the correlation time of the fluctuations that result in linewidth broadening. Longer correlation time results in an autocorrelation function decaying more slowly with time [17]. Equation (3.56) is an exponentially decaying cosine wave. The smaller the loss, the longer its duration. Furthermore, for $\tau = 0$ we obtain $R_{VV}(0) = A_1^2/2$ which is the carrier power. The autocorrelation function is also a measure of the self-resemblance of the signal with its displaced version [20]. Equation (3.56) shows that the carrier self-similarity persists more, the smaller the diffusion constant D is. As the amplitude of the oscillator carrier is constant due to energy balance, the decay with time of self-resemblance suggests an oscillator carrier with diverging/diffusing phase.

The Fourier transform of (3.56) is calculated with the aid of the Fourier Transform pairs

$$\pi[\delta(\omega_o + \omega) + \delta(\omega_o - \omega)] \leftrightarrow \cos(\omega_o t) \qquad (3.57)$$

$$\frac{2\alpha}{\omega^2 + \alpha^2} \leftrightarrow e^{-\alpha|t|} \qquad (3.58)$$

and the modulation property of the Fourier Transform

$$\frac{1}{2\pi}(F_1(\omega) * F_2(\omega)) \leftrightarrow f_1(t) \cdot f_2(t) \qquad (3.59)$$

where star denotes convolution. Therefore we obtain

$$S_\theta(\omega) = \frac{A_1^2}{2} \cdot \frac{1}{2\pi} \cdot \left[\frac{2D}{\omega^2 + D^2} * \pi[\delta(\omega_o + \omega) + \delta(\omega_o - \omega)] \right] \qquad (3.60)$$

which is further written as

$$S_\theta(\omega) = \frac{A_1^2}{2} \cdot \left[\frac{D}{(\omega_o + \omega)^2 + D^2} + \frac{D}{(\omega_o - \omega)^2 + D^2} \right] \qquad (3.61)$$

The first term in the sum corresponds to negative frequencies, while the second term corresponds to the positive part of the spectrum. Substituting $D = \beta_\theta/2$ from (3.47) results in

$$S_\theta(\omega) = \frac{A_1^2 \beta_\theta}{4} \cdot \left[\frac{1}{(\omega_o + \omega)^2 + (\beta_\theta/2)^2} + \frac{1}{(\omega_o - \omega)^2 + (\beta_\theta/2)^2} \right] \qquad (3.62)$$

Equation (3.62) is the same as (3.29), showing that the dissipation-based approach presented in Sect. 3.4 gives the same result as the fluctuation-based approach presented in this section. This is a result of the fluctuation-dissipation theorem that relates the autocorrelation function of the fluctuations to the dissipation response function (see Sect. 3.2).

The connection between fluctuation and dissipation is further highlighted by considering the energy loss in a weakly damped oscillator. As was shown in (1.28), the oscillator energy diminishes exponentially with time giving

$$\frac{|\Delta E|}{\Delta t} = \beta_\theta E \tag{3.63}$$

Equations (3.41), (3.47), and (3.63) thus give

$$\frac{|\Delta E|}{E} = \beta_\theta\, t = 2\,D\,t = \sigma_\theta{}^2 \tag{3.64}$$

The phase variance is proportional to energy loss and inversely proportional to the oscillator energy. Minimizing phase noise, therefore, involves minimizing the tank loss and maximizing the energy stored in the tank.

3.6 Stochastic Differential Equation for Phase

In this section, we look into the fluctuation approach more formally. We derive and solve the stochastic differential equation that describes the time evolution of the random phase and prove that the oscillator phase is a diffusion process. We start from the equivalent network shown in Fig. 3.4. Assuming that over an oscillation cycle, the nonlinear active element $i(v)$ cancels out the action of the resistor R, the differential equation describing the circuit becomes

$$n(t) = C\frac{dv(t)}{dt} + \frac{1}{L}\int v(t)dt \tag{3.65}$$

where we have used $n(t)$ instead of $i_n(t)$ to denote the noise current. Setting $x(t) = \int v(t)\,dt$, $\dot{x}(t) = v(t)$ and $\ddot{x}(t) = \dot{v}(t)$, Eq. (3.65) becomes

$$\ddot{x}_t + \omega_o^2 x_t = \frac{n_t}{C} \tag{3.66}$$

The subscript t denotes time dependency, while overdot stands for time derivative. This is the equation of a simple harmonic oscillator driven by noise. We write (3.66) as two coupled first-order equations

$$\dot{x}_t = y_t \tag{3.67}$$

$$\dot{y}_t = -\omega_o^2 x_t + \frac{n_t}{C} \tag{3.68}$$

Since we are interested in phase, we switch from Cartesian coordinates x_t, y_t to polar coordinates r_t, θ_t by using the transformation equations

$$r_t^2 = (\omega_o x_t)^2 + y_t^2 \tag{3.69}$$

$$\tan(\theta_t) = \frac{y_t}{\omega_o x_t} \tag{3.70}$$

The energy balance between $i(v)$ and R sets the oscillation amplitude to A_1 as discussed in Chap. 2. Thus, we discard the amplitude equation and regard r_t as constant and equal to A_1[1]. To obtain the differential equation for the phase, we differentiate (3.70) with respect to time

$$\frac{\dot{\theta}_t}{\cos^2(\theta_t)} = \frac{1}{\omega_o} \frac{x_t \dot{y}_t - y_t \dot{x}_t}{x_t^2} \tag{3.71}$$

Substituting for $\dot{x}_t$, $\dot{y}_t$ from Eqs. (3.67) and (3.68) and using $\omega_o x_t = A_1 \cos(\theta_t)$ and $y_t = A_1 \sin(\theta_t)$ we obtain a first-order stochastic differential equation for the oscillator phase

$$\dot{\theta}_t = -\omega_o + \frac{1}{CA_1} \cos(\theta_t) n_t \tag{3.72}$$

which we further write in the form

$$d\theta_t = -\omega_o dt + \frac{1}{CA_1} \cos(\theta_t) dn_t \tag{3.73}$$

where dn_t[2] is shorthand for $n_{t+dt} - n_t$ [16]. The first term on the right side of (3.73) represents the deterministic part of the phase, while the second term represents the stochastic part. We subsequently express dn_t as a Wiener process [16, 21]

$$dn_t = n_{t+dt} - n_t = \sqrt{\delta^2 \, dt} \cdot N_t^{t+dt}(0, 1) \tag{3.74}$$

where

$$\delta^2 = S_{i_n}(f) = \frac{\overline{i_n^2}}{df} = \frac{2kT}{R} \tag{3.75}$$

---

[1] Appendix A treats amplitude noise.
[2] dn_t has units of electric charge.

In (3.74), $N_t^{t+dt}(0, 1)$ denotes a normal distribution with zero mean and unit standard deviation associated explicitly with the interval $(t, t + dt)$. The interpretation of (3.74) is that when the Wiener process realizes a value n_t at time t, its realization n_{t+dt} at time $t + dt$ is given by n_t plus the product of $\sqrt{\delta^2 dt}$ times the realization of a normal distribution with zero mean and unit standard deviation. Normals associated with disjunct time intervals are statistically independent, meaning that the correlation time of the fluctuating current is much shorter than the characteristic time of the oscillator. Seen from a numerical perspective, (3.74) constitutes a Monte Carlo simulation.

In order to solve the stochastic differential equation (3.73), we need to determine the mean value $\overline{\theta_t}$ and the variance $\sigma_\theta{}^2$. First we express (3.73) in the form

$$d\theta_t = -\omega_o \, dt + \epsilon \cos(\theta_t) dW_t \tag{3.76}$$

where dW_t denotes a Wiener process with zero mean and unit standard deviation

$$dW_t = \sqrt{dt} \cdot N_t^{t+dt}(0, 1) \tag{3.77}$$

and term ϵ is given by

$$\epsilon = \frac{\delta}{CA_1} \tag{3.78}$$

Taking the average of (3.76) and noting that $\overline{dW_t} = 0$ gives

$$d(\overline{\theta_t}) = -\omega_o \, dt \tag{3.79}$$

which with zero initial conditions results in

$$\overline{\theta_t} = -\omega_o \, t \tag{3.80}$$

This is the expected deterministic solution for the noiseless simple harmonic oscillator phase. The next step is determining $\overline{\theta_t{}^2}$. For this, we use

$$d(\theta_t^2) = (\theta_t + d\theta_t)^2 - \theta_t^2 \tag{3.81}$$

Therefore

$$d(\theta_t^2) = 2\theta_t d\theta_t + (d\theta_t)^2 \tag{3.82}$$

Substituting (3.76) into (3.82) gives

$$d(\theta_t^2) = -2\theta_t \omega_o dt + 2\theta_t \epsilon \cos(\theta_t) dW_t + (d\theta_t)^2 \tag{3.83}$$

taking the average gives

$$d(\overline{\theta_t^2}) = -2\,\overline{\theta_t}\,\omega_o\,dt + 2\overline{\theta_t}\,\epsilon\,\overline{\cos(\theta_t)\,dW_t} + \overline{(d\theta_t)^2} \tag{3.84}$$

substituting for $\overline{\theta_t}$ from (3.80) results in

$$d(\overline{\theta_t^2}) = 2\,\omega_o{}^2 t\,dt - 2\,\omega\,t\,\epsilon\,\overline{\cos(\theta_t)\,dW_t} + \overline{(d\theta_t)^2} \tag{3.85}$$

Noting that $\overline{dW_t} = 0$ simplifies (3.85) to

$$d(\overline{\theta_t^2}) = 2\,\omega_o{}^2 t\,dt + \overline{(d\theta_t)^2} \tag{3.86}$$

To estimate the second term $\overline{(d\theta_t)^2}$, we square and average (3.76)

$$\overline{(d\theta_t)^2} = \omega_o{}^2\,dt^2 + \epsilon^2\,\overline{\cos^2(\theta_t)\,(dW_t)^2} - 2\,\omega_o\,dt\,\epsilon\,\overline{\cos(\theta_t)\,dW_t} \tag{3.87}$$

In the limit $dt \to 0$, the second term dominates as $\overline{(dW_t)^2} = dt$. Since $\overline{\cos^2(\theta_t)} = 1/2$, (3.87) becomes

$$\overline{(d\theta_t)^2} = \frac{1}{2}\,\epsilon^2\,dt \tag{3.88}$$

Substituting (3.88) into (3.86) gives

$$d(\overline{\theta_t^2}) = 2\,\omega_o{}^2 t\,dt + \frac{1}{2}\,\epsilon^2\,dt \tag{3.89}$$

which, with zero initial conditions, results in

$$\overline{\theta_t^2} = \omega_o{}^2\,t^2 + \frac{1}{2}\,\epsilon^2\,t \tag{3.90}$$

The phase variance can be expressed by

$$\sigma_\theta{}^2 = \overline{\theta_t^2} - \overline{\theta_t}^2 \tag{3.91}$$

Substituting (3.80) and (3.90) into (3.91) gives

$$\sigma_\theta{}^2 = \frac{1}{2}\,\epsilon^2\,t = \frac{\delta^2}{2C^2 A_1{}^2}\,t \tag{3.92}$$

Since we have assumed that θ is the sum of many uncorrelated Gaussian variables, it is Gaussian. Its probability density function is therefore given by

$$P(\theta, t) = \frac{1}{\sqrt{2\pi\sigma_\theta^2}} e^{-\frac{\theta^2}{2\sigma_\theta^2}} \tag{3.93}$$

which is the solution to the diffusion equation

$$\frac{\partial P}{\partial t} = \left(\frac{\sigma_\theta^2}{2t}\right) \cdot \frac{\partial^2 P}{\partial^2 \theta} \tag{3.94}$$

as can be verified by substituting (3.93) into (3.94). The diffusion equation (3.94) is mathematically equivalent to the stochastic differential equation (3.73). The former governs the probability density of the phase, while the latter governs the time evolution of the random phase [16]. Using (3.92) the diffusion constant becomes

$$D = \frac{\sigma_\theta^2}{2t} \Rightarrow D = \frac{\delta^2}{4C^2 A_1^2} \tag{3.95}$$

and finally substituting $\delta^2 = 2kT/R$ in (3.95) gives for the phase diffusion constant

$$D = \frac{kT}{2RC^2 A_1^2} \tag{3.96}$$

which is the same as (3.45).

Figure 3.8 shows hundred different runs of the time evolution of the random phase, while Fig. 3.9 shows the probability density for the random phase at 1 μs, 10 μs, 100 μs, and 1 ms, respectively. Both Figures show clearly the spreading out of the random phase with time. In both cases we have set $C = 1$ pF, $L = 1$ nH, $R = 500$, and $A_1 = 1$V.

Fig. 3.8 Time evolution of random phase of a self-sustained oscillator with $C = 1$ pF, $L = 1$ nH, $R = 500$ and $A_1 = 1$ V

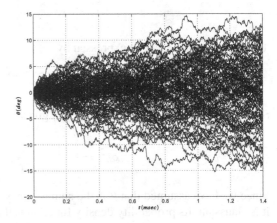

Fig. 3.9 The probability
density for the random phase
of a self-sustained oscillator
with $C = 1$ pF, $L = 1$ nH,
$R = 500$ and $A_1 = 1$ V at
1 μs, 10 μs, 100 μs, and
1 ms, respectively

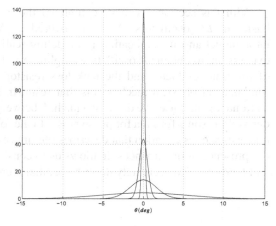

Fig. 3.10 Phase noise plot of
the self-sustained oscillator
depicted in Fig. 3.4, with
$C = 1$ pF, $L = 1$ nH,
$R_p = 500$ Ω, and $A_1 = 1$ V.
The x-axis shows the
frequency offset from the
carrier frequency. Circles:
simulated phase noise.
Continuous line: estimated
Phase Noise according to
(3.97)

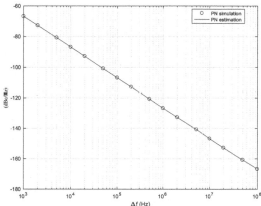

3.7 Comparison with Simulation

In this chapter, we have used the two approaches of dissipation and fluctuation
to derive the fundamental Equation that gives the phase noise in a self-sustained
oscillator due to tank loss. This is Leeson's formula (3.32) repeated below

$$PN = \frac{kT}{R\,C^2\,A_1{}^2\,(\Delta\omega)^2} \tag{3.97}$$

Figure 3.10 shows the phase noise of the self-sustained oscillator depicted in Fig. 3.4
against frequency offset from the carrier frequency. We have used $C = 1$ pF, $L = 1$ nH, $R_p = 500$ Ω, and $A_1 = 1$ V. The continuous curve is the prediction of the
Eq. (3.97), while the circles correspond to the simulation. Exact agreement between
simulation and (3.97) is observed.

With this section, we have completed the three introductory chapters on the basics of *LC* oscillators. We have covered the Van der Pol self-sustained oscillator model and its application in modeling realistic oscillators. We have derived expressions to obtain the oscillation amplitude and its harmonics from the properties of the transconductor and the tank loss resistor. Subsequently, we discussed the fundamental physical mechanisms that result in lineshape broadening due to the stationary tank noise. We discussed the link between fluctuations and dissipation and derived Leeson's formula for phase noise. In the following chapter, we introduce the phase dynamics equation that describes the time evolution of the oscillator phase in the presence of disturbances. It allows us to consider cyclostationary noise sources, and lends itself to accurately describe oscillator entrainment and pulling.

Chapter 4
Thermal Noise in LC Oscillators

4.1 Introduction

This chapter delves further into methods for analyzing phase noise in LC-tuned oscillators. We start by establishing the phase dynamics equation that encapsulates the sensitivity of the oscillator phase to disturbances. We then apply it to two cases of practical interest. The first one concerns the tank's thermal noise. The second one examines the oscillator's phase noise due to the transconductor thermal noise.

4.2 The Phase Dynamics Equation

Figure 4.1 depicts a model for a self-sustained oscillator. The LC tank defines the oscillation frequency $\omega_o = 1/\sqrt{LC}$, while resistor R accounts for losses. Over the oscillation period, the energy balance between the nonlinear current source $i(v)$ and the resistor R maintains the oscillation and defines the oscillation amplitude as discussed in Chap. 2. Current source i_{inj} disturbs the oscillator motion. To estimate the oscillator amplitude and phase deviations from their unperturbed values, we simplify the oscillator model as is shown in the lower part of Fig. 4.1. Essentially, we assume that R and $i(v)$ cancel out, and for frequencies slightly away from ω_o (see Sect. 3.4), the oscillator behavior can be approximated by a lossless LC network. We also assume that the injected current is sufficiently small.

We express the oscillation carrier as

$$v = A_1 \cos(\omega_o t) \tag{4.1}$$

while the tank current i is given by

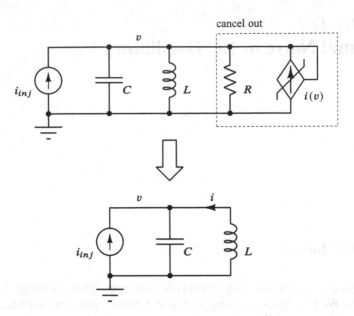

Fig. 4.1 Model for a self-sustained oscillator and its simplification for noise calculations

$$i = C\frac{dv}{dt} = -A_1 C\omega_o \sin(\omega_o t) \tag{4.2}$$

Let us assume that over time Δt current source i_{inj} injects charge Δq into the tank. The charge flows into the tank capacitor altering its voltage according to

$$v = A_1 \cos(\omega_o t) + \frac{\Delta q}{C} = (A_1 + \Delta A_1)\cos(\omega_o t + \Delta\phi) \tag{4.3}$$

where ΔA_1 and $\Delta\phi$ denote the amplitude and phase deviations. The inductor resists any current change, so the tank current remains constant, giving

$$i = -A_1 C\omega_o \sin(\omega_o t) = -(A_1 + \Delta A_1)C\omega_o \sin(\omega_o t + \Delta\phi) \tag{4.4}$$

Putting Eqs. (4.3) and (4.4) together results in the system of equations

$$\Delta A_1 \cos(\omega_o t) - A_1 \sin(\omega_o t)\Delta\phi = \frac{\Delta q}{C} \tag{4.5}$$

$$\Delta A_1 \sin(\omega_o t) + A_1 \cos(\omega_o t)\Delta\phi = 0 \tag{4.6}$$

where we used $\sin(\Delta\phi) \approx \Delta\phi$, $\cos(\Delta\phi) \approx 1$, and $A_1 + \Delta A_1 \approx A_1$. Solving for the amplitude and phase deviations, we obtain [22]

Fig. 4.2 Geometrical interpretation of amplitude and phase deviations

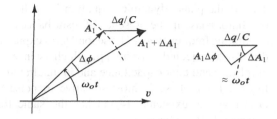

$$\Delta A_1 = \frac{\Delta q}{C} \cos(\omega_o t) \qquad (4.7)$$

$$\Delta \phi = -\frac{\Delta q}{A_1 C} \sin(\omega_o t) \qquad (4.8)$$

These are the same as (3.17) and (3.18).[1] As the carrier is a cosine wave, the sin term in (4.8) shows that phase is most sensitive to disturbances when the carrier crosses zero, while the cos term in (4.7) shows that amplitude is most sensitive to disturbances when the carrier is at its peak values. A geometrical interpretation of (4.7) and (4.8) is depicted in Fig. 4.2. The carrier is a rotating vector of length A_1, traversing the oscillator limit cycle counterclockwise. The injection of charge Δq results in amplitude and phase deviations ΔA_1 and $\Delta \phi$, respectively. From the right triangle on the right part of the figure, we can directly obtain Eqs. (4.7) and (4.8).

The self-sustained oscillator's inherent amplitude control eventually suppresses the amplitude deviations. On the other hand, phase deviations persist. Therefore, we will concentrate on (4.8). Dividing (4.8) by Δt and taking the limit as $\Delta t \to 0$, we obtain the differential equation

$$\frac{d\phi}{dt} = \frac{dq/dt}{A_1 C} \sin(\omega_o t) \qquad (4.9)$$

The term dq/dt is the externally injected current i_{inj} responsible for the phase deviation. Equation (4.9) thus takes the form

$$\frac{d\phi}{dt} = \frac{i_{inj}}{A_1 C} \sin(\omega_o t) \qquad (4.10)$$

For the total carrier phase, we can write

$$\frac{d\phi_{tot}}{dt} = \omega_o + \frac{i_{inj}}{A_1 C} \sin(\omega_o t) \qquad (4.11)$$

---

[1] The apparent difference between (4.7), (4.8) and (3.17) and (3.18) is that in this chapter we consider a cosine-wave carrier, while in the previous chapter, we considered a sine-wave carrier.

This is the phase dynamics equation [13]. It describes the time evolution of the oscillator phase in the presence of disturbances. The first term is due to the average oscillation frequency ω_o. The second term expresses the oscillator phase disturbance due to the injected current i_{inj}, which is inversely proportional to the oscillator amplitude and tank capacitance and in quadrature to the oscillator carrier. As noted, the oscillator phase is most sensitive to noise injection around the carrier zero crossings. As expected, (4.11) has the same form as the stochastic differential equation for the phase (3.72).

The phase dynamics equation forms the basis for phase noise analysis in this and the next chapter. We will apply it to several cases of practical interest, including tank thermal noise, transconductor thermal and flicker noise, power supply, and bias low-frequency noise. Furthermore, the phase dynamics equation can analyze phenomena modeled as a current injection across the oscillator tank, such as oscillator entrainment and pulling.

4.3 Noise to Phase Noise Conversion

From our discussion on the driven harmonic oscillator in Sect. 1.6, we expect that the oscillator is sensitive only to disturbances introduced close to its natural frequency ω_o. Therefore, only noise components near ω_o affect the oscillator phase. This is the case if stationary noise is injected across the oscillator tank. A typical example of stationary noise is tank thermal noise. As we are interested in spot noise (per Hz), in the following, we model stationary noise current sources i_{n_s} at frequency ω by

$$i_{n_s} = i_{n_o} \cos(\omega t + \psi_n) \tag{4.12}$$

i_{n_o} is in $A/\sqrt{Hz}$ so that $i_{n_o}^2/2$ corresponds to the noise current density in A^2/Hz. Phase ψ_n is uniformly distributed in $[0, 2\pi]$.

Noise from active devices depends on their terminal voltages and currents. As these vary over the oscillator period, so do the statistics of their noise densities. Such noise sources are called cyclostationary to emphasize that their noise is periodically injected into the tank. Therefore, noise emanating from the oscillator transconductor, the oscillator bias, and the oscillator supply circuitry is cyclostationary. We model cyclostationary noise current sources $i_{n_{cs}}$ by

$$\overline{i_{n_{cs}}^2}/df = \left[\overline{i_{n_s}^2}/df\right] \cdot G(t) \tag{4.13}$$

where i_{n_s} is stationary and $G(t)$ is periodic with period $T = 2\pi/\omega_o$.

The effect of $G(t)$ in (4.13) is to translate noise from the oscillator harmonics to ω_o. Effectively, each harmonic of $G(t)$ at $n\omega_o$ picks up noise from $(n - 1)\omega_o$ and $(n + 1)\omega_o$ and sends it to ω_o. When modeling this process, we need to keep track of noise correlations. The phase dynamics equation governs the conversion of

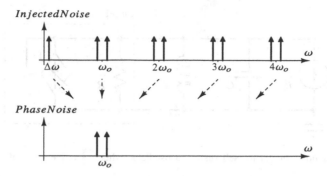

Fig. 4.3 The periodic function G(t) translates noise from $\Delta\omega$ and $n\omega_o \pm \Delta\omega$ with $n = 1, 2, \ldots$ to $\omega_o \pm \Delta\omega$. The phase dynamics equation governs the conversion of the translated noise to phase noise

the translated noise to phase noise. The overall process is that noise from all the oscillator's harmonics affects the phase noise around the oscillator carrier, as shown schematically in Fig. 4.3.

Therefore, noise to phase noise conversion is a two-step process. The first step is due to cyclostationarity and results in noise translations from the oscillator harmonics to ω_o. These frequency translations are generally unavoidable, as they are the effect of the operation of the oscillator circuit. The second step involves the conversion of the translated noise to phase noise. The phase dynamics equation governs this step.

4.4 Tank Noise and Leeson's Equation

Let us consider tank noise as the first application of the phase dynamics equation. The aim is to apply the phase dynamics equation to derive Leeson's formula (3.97). We work with the model shown in Fig. 4.4. The energy balance between the nonlinear current source $i(v)$ and the tank loss resistor R maintains the oscillation and defines the oscillation amplitude. For noise calculations, we may assume that $i(v)$ and R cancel out over the oscillator period, and we disregard them.

The injected tank noise is stationary. Therefore, only noise near ω_o matters. We express the tank thermal noise components at frequency offsets $\pm\Delta\omega$ from the carrier as

$$i_n = i_o \cos\left[(\omega_o \pm \Delta\omega)t + \psi_\pm\right] \tag{4.14}$$

where noise components both above and below the carrier are accounted for. Following the convention we introduced in (4.12), $i_o^2/2 = 4kT/R$, while $\psi_\pm$ are uncorrelated and uniformly distributed in $[0, 2\pi]$. The phase dynamics equa-

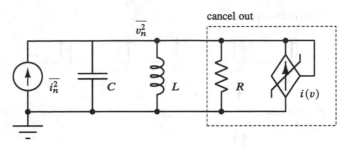

Fig. 4.4 Tank noise model for self-sustained oscillator, where $\overline{i_n^2}$ is the noise of the tank loss resistor R

tion (4.11) becomes

$$\frac{d\phi_{tot}}{dt} = \omega_o + \frac{i_o}{A_1 C} \cos\left[(\omega_o \pm \Delta\omega)t + \psi_\pm\right] \sin(\omega_o t) \tag{4.15}$$

Expanding (4.15) and keeping only the low-frequency terms gives

$$\frac{d\phi_{tot}}{dt} = \omega_o - \frac{i_o}{2A_1 C} \sin(\pm\Delta\omega\, t + \psi_\pm) \tag{4.16}$$

We subsequently integrate (4.16) to obtain the oscillator phase as

$$\phi_{tot} = \omega_o t$$

$$+ \frac{i_o}{2A_1 C \Delta\omega}(-\cos\psi_- + \cos\psi_+)\cos(\Delta\omega\, t)$$

$$+ \frac{i_o}{2A_1 C \Delta\omega}(-\sin\psi_- - \sin\psi_+)\sin(\Delta\omega\, t) \tag{4.17}$$

The oscillator carrier can now be expressed as $v = A_1 \cos(\phi_{tot})$. Assuming that the noise terms are small, we obtain

$$v = A_1 \cos(\omega_o t)$$

$$+ \frac{i_o}{4C \Delta\omega} \sin\left[(\omega_o + \Delta\omega)t + \psi_+\right]$$

$$- \frac{i_o}{4C \Delta\omega} \sin\left[(\omega_o + \Delta\omega)t - \psi_-\right] \tag{4.18}$$

We have kept only the upper sideband terms because we are interested in single-sided phase noise. This further becomes

$$v = A_1 \cos(\omega_o t)$$

$$+ \frac{i_o}{4C\Delta\omega} (\sin \psi_- + \sin \psi_+) \cos \left[(\omega_o + \Delta\omega)t \right]$$

$$+ \frac{i_o}{4C\Delta\omega} (- \cos \psi_- + \cos \psi_+) \sin \left[(\omega_o + \Delta\omega)t \right] \qquad (4.19)$$

The carrier power is $P_{carrier} = A_1^2/2$, while for the noise power we obtain from (4.19)

$$P_{noise} = \frac{i_o^2/2}{8C^2\Delta\omega^2} \qquad (4.20)$$

where we took advantage of the fact that ψ_- and ψ_+ are uncorrelated. The final step is to express the single-sided phase noise as the noise power ratio to the carrier power. This gives

$$PN_{tank} = \frac{P_{noise}}{P_{carrier}} = \frac{kT}{A_1^2 RC^2(\Delta\omega)^2} \qquad (4.21)$$

where we used that $i_o^2/2 = 4kT/R$.

Equation (4.21) is, of course, Leeson's Eq. (3.97) [18]. Utilizing the phase dynamics equation produced the same result as the dissipation and fluctuation approaches presented in the previous chapter. However, as we will immediately demonstrate, this method bears the crucial advantage that it can be easily extended to tackle cyclostationary noise sources. It is, therefore, more powerful, and we will utilize it extensively in the following.

4.5 Transconductor Modeling

A typical N-only transconductor is shown in Fig. 4.5. V_d is the differential voltage across the transconductor. The individual device currents I_1 and I_2 can be expressed in terms of the differential and common-mode currents I_d and I_{cm}, respectively, as

$$I_{1,2} = \frac{I_{cm}}{2} \pm I_d \qquad (4.22)$$

The differential current I_d flows into the tank differential impedance. In contrast, the common-mode current I_{cm} flows from the supply via the tank common-mode impedance into the transconductor and terminates to the ground. We model the differential current as

$$I_d = g_1 V_d + g_3 V_d^3 \qquad (4.23)$$

Fig. 4.5 N-only
transconductor connected
across a tank circuit

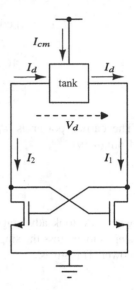

We expect that g_1 and g_3 have opposite signs, so the transconductor exhibits limiting behavior. The oscillation amplitude is given by (2.69) as

$$A_1 = \sqrt{\frac{4(g_1 R - 1)}{3 g_3 R}} \qquad (4.24)$$

where R is the tank loss resistor.

The common-mode current is modeled as

$$I_{cm} = I_o + g_2 V_d{}^2 + g_4 V_d{}^4 + g_6 V_d{}^6 \qquad (4.25)$$

where I_o is the transconductor quiescent current when no differential swing appears across it. The common-mode current does not affect the oscillation amplitude. It, however, contributes to translating the transconductor thermal noise from the oscillator harmonics to the fundamental. Figures 4.6 and 4.7 present a comparison between simulated currents and the respective models obtained by fitting Eqs. (4.23) and (4.25) to the simulated data.

4.6 Phase Noise Due to Transconductor Thermal Noise

In the following, we will assume that the channel thermal noise density $\overline{i_{th}{}^2}/df$ of a CMOS device is proportional to the square root of the drain current I_D. This is expected from the well-known expression $\overline{i_{th}{}^2}/df = 4kT\gamma g_m$, where the device transconductance $g_m = \sqrt{2\mu C_{ox}(W/L)I_D}$ is proportional to the square root of the

Fig. 4.6 Continuous curves: Simulated I_d and I_{cm}. Dashed curves: Models of I_d and I_{cm} derived using Eqs. (4.23) and (4.25), respectively. Both are plotted against the oscillator period in rad. The simulated oscillator is shown in Fig. 4.8

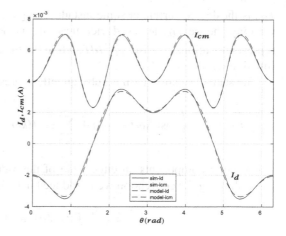

Fig. 4.7 Continuous curves: Simulated I_1 and I_2. Dashed curves: Models of I_1 and I_2 derived using Eqs. (4.22), (4.23), and (4.25). Both are plotted against the oscillator period in rad. The simulated oscillator is shown in Fig. 4.8

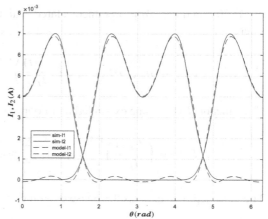

drain current I_D in saturation. W and L are the device width and length, while μ and C_{ox} are the mobility and gate oxide capacitance. Term γ is a fabrication process parameter. Under the above assumption, the cyclostationary channel thermal noise can be expressed as

$$\overline{i_{th_{cs}}^2}/df = 4kT\gamma\sqrt{2\mu C_{ox}(W/L)}\sqrt{I_D(t)} \qquad (4.26)$$

where $I_D(t)$ is positive.

Equation (4.26) can be further written as

$$\overline{i_{th_{cs}}^2}/df = \left[4kT\gamma\sqrt{2\mu C_{ox}(W/L)I_{Do}}\right]\sqrt{\frac{I_D(t)}{I_{Do}}} = \left[\overline{i_{th_s}^2}/df\right] \cdot G(t) \qquad (4.27)$$

I_{Do} is the device current at some instant during the oscillator period. For example, we may choose I_{Do} as the device bias current. Per (4.13), the term in brackets represents stationary noise $\overline{i_{th_s}^2}/df$, while the time-dependent term is the periodic function $G(t)$.

The stationary noise term is subsequently expressed as in (4.14) by

$$i_{th_s} = i_o \cdot \left(\cos(\Delta\omega t + \psi_0) + \sum_{k=1}^{\infty} \cos((k\omega_o \pm \Delta\omega)t + \psi_{k\pm}) \right) \qquad (4.28)$$

where noise components on either side of the oscillator harmonics are accounted for. Following the convention we introduced in (4.12), $i_o^2/2$ is given by

$$i_o^2/2 = 4kT\gamma\sqrt{2\mu C_{ox}(W/L)I_{Do}} \qquad (4.29)$$

in A^2/Hz and $\psi_0, \psi_{k\pm}$ are uncorrelated and uniformly distributed in $[0, 2\pi]$. Therefore,

$$i_{th_{cs}} = i_{th_s}\sqrt{G(t)} = i_{th_s}\left(\frac{I_D(t)}{I_{Do}}\right)^{1/4} = i_{th_s} P(t) \qquad (4.30)$$

The function $P(t)$ can be expanded in a Fourier series

$$P(t) = \left(\frac{I_D(t)}{I_{Do}}\right)^{1/4} = p_0 + \sum_{l=1}^{\infty} p_l \cos(l\omega_o + \theta_l) \qquad (4.31)$$

The next step is to express the phase dynamics equation (4.11)

$$\frac{d\phi_{tot}}{dt} = \omega_o + \frac{i_{inj}}{A_1 C}\sin(\omega_o t) \qquad (4.32)$$

where the injected current into the tank $i_{inj} = i_{th_s} P(t)$ is the product of (4.28) and (4.31)

$$i_{inj} = i_o \cdot \left(\cos(\Delta\omega t + \psi_0) + \sum_{k=1}^{\infty} \cos((k\omega_o \pm \Delta\omega)t + \psi_{k\pm}) \right)$$

$$\cdot \left(p_0 + \sum_{l=1}^{\infty} p_l \cos(l\omega_o + \theta_l) \right) \qquad (4.33)$$

To keep the analysis tractable, we discard all harmonics above the third in (4.31). Therefore, noise only up to the fourth harmonic is considered. This simplification is justified by comparing the predicted phase noise to simulations (see Sect. 4.8).

Furthermore, because of the functional relationship between $I_D(t)$ and $P(t)$ given in (4.30), the harmonics of $P(t)$ are either in-phase or anti-phase with the harmonics of the transconductor current. This allows us to discard all the phase terms θ_l in (4.33). With these simplifications, the phase dynamics equation takes the form

$$\frac{d\phi_{tot}}{dt} = \omega_o + \frac{i_o}{A_1 C}$$

$$\cdot \left(\cos(\Delta\omega\, t + \psi_0) + \sum_{k=1}^{4} \cos((k\omega_o \pm \Delta\omega)t + \psi_{k\pm}) \right)$$

$$\cdot \left(\sum_{l=0}^{3} p_l \cos(l\omega_o) \right) \cdot \sin(\omega_o t) \tag{4.34}$$

Expanding (4.34) and following a procedure similar to the one detailed in Sect. 4.4 give for the transconductor upper sideband noise power the following expression:

$$P_{noise} = 2 \cdot \frac{(i_o{}^2/2)}{4C^2 \Delta\omega^2} \cdot \left(\frac{(p_0 + p_2)^2 + p_0^2}{16} + \frac{(p_1 + p_3)^2 + p_3^2}{32} \right) \tag{4.35}$$

The factor of two in (4.35) accounts for the two transconductor devices adding noise simultaneously, while the factor of four in the denominator is introduced because the transconductor devices inject noise only on one side of the tank. The term in the parenthesis

$$F_{cs} = \frac{(p_0 + p_2)^2 + p_0^2}{16} + \frac{(p_1 + p_3)^2 + p_3^2}{32} \tag{4.36}$$

encapsulates the effect of the cyclostationary nature of the transconductor noise. It captures the translation of noise from the oscillator harmonics to the oscillator carrier frequency, while, at the same time, it accounts for noise correlations. Terms p_0, p_1, p_2, etc., are the harmonics of the function $P(t)$ defined in (4.30)

$$P(t) = \left(\frac{I_D(t)}{I_{Do}} \right)^{1/4} \tag{4.37}$$

The oscillator carrier power is $P_{carrier} = A_1^2/2$, so the single-sided phase noise can be readily expressed as

$$PN_{Gm} = \frac{P_{noise}}{P_{carrier}} = \frac{(i_o{}^2/2)F_{cs}}{A_1{}^2 C^2 (\Delta\omega)^2} \tag{4.38}$$

Equation (4.38) is remarkably simple. On the numerator, we have the stationary noise of a single transconductor device at its bias point as expressed in (4.29). This is multiplied by the cyclostationary factor F_{cs} given in (4.36). On the denominator, we have the standard terms: the oscillation amplitude squared, the tank capacitance squared, and the square of the offset frequency from the carrier.

4.7 Excess Noise Factor

The total phase noise taking into account both tank loss noise and transconductor thermal noise is simply the sum of (4.21) and (4.38)

$$PN_{tot} = PN_{tank} + PN_{Gm} \qquad (4.39)$$

It is instructive to manipulate (4.39) to obtain an expression for the excess noise factor (ENF). ENF measures how much higher the total phase noise is than the phase noise due to tank loss alone. Using (4.21) and (4.38), we obtain

$$PN_{tot} = \frac{kT}{A_1^2 RC^2 \Delta\omega^2} \cdot \left[1 + 4\frac{(i_o^2/2)}{(4kT/R)} F_{cs} \right] \qquad (4.40)$$

The excess noise factor thus becomes

$$ENF = 1 + 4\frac{(i_o^2/2)}{(4kT/R)} F_{cs} \qquad (4.41)$$

Finally, using $i_o^2/2 = 4kT\gamma g_m$, where g_m is the transconductance of each device at the bias point, we obtain

$$ENF = 1 + 4\gamma \cdot g_m R \cdot F_{cs} \qquad (4.42)$$

Remembering the discussion in Chap. 2, the term $g_m R$ is larger than unity to guarantee startup, while the term γ depends on the fabrication process. The term F_{cs} given in (4.36) captures the cyclostationary nature of the transconductor noise.

4.8 Comparison with Simulations

This section compares (4.38) with simulation. We seek to verify that (4.38) accurately captures the physical processes that govern the conversion of transconductor thermal noise into phase noise around the oscillator carrier. For this, we utilize a VerilogA MOS device model that represents the drain current using a single expression that is continuous and valid in all regions of operation [14].

Fig. 4.8 Schematic of the
simulated oscillator. The
250 Ω resistor models tank
loss and is considered
noiseless. Only thermal noise
from the transconductor
active devices is included

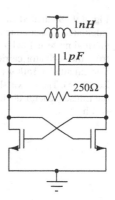

$$I_D = 2\frac{W}{L}K_p n {\phi_t}^2$$

$$\cdot\left[ln^2\left(1+e^{\frac{V_{GS}-V_T}{2n\phi_t}}\right) - ln^2\left(1+e^{\frac{V_{GS}-V_T-nV_{DS}}{2n\phi_t}}\right)\right] \qquad (4.43)$$

W, L, and K_p are the device width, length, and transconductance parameter,
respectively. V_{GS}, V_{DS} are the gate–source and drain–source voltages. ϕ_t is the
thermal voltage, V_T is the threshold voltage, and n is the subthreshold slope. The
device thermal noise is modeled according to (4.26).

Figure 4.8 shows the oscillator schematic used for the comparison. The tank
components are chosen so that the oscillator carrier is centered at 5 GHz. The 250 Ω
resistor models tank loss and is considered noiseless. For the devices, we have used
$L = 100$ nm, $K_p = 200\,\mu A/V^2$, $V_T = 0.35$ V, and $n = 1$. Only thermal noise from
the transconductor active devices is included. Figure 4.9 shows the transconductor
thermal noise at 1 MHz offset from the oscillator carrier. The calculation follows
Eq. (4.35). Figure 4.10 shows the phase noise due to the transconductor thermal
noise at 1 MHz offset from the oscillator carrier. The calculation follows Eq. (4.38).
A very good agreement between simulation and calculation is observed in both
figures.

As the device width increases, the simulated and calculated curves deviate. This
is because the oscillator amplitude increases with the device width, resulting in
transconductor thermal noise components beyond the first four harmonics finding
their way to the oscillator carrier. As mentioned before, Eq. (4.35) accounts for noise
only up to the fourth harmonic and thus slightly underestimates the total translated
noise. The underestimation increases with the device width. Figure 4.11 depicts
the calculated cyclostationary factor F_{cs} against the width of the transconductor
devices.

This chapter delved into the physical mechanisms that govern noise to phase
noise conversion in oscillators. We have highlighted the role of cyclostationarity in
translating noise from the oscillator's harmonics to the oscillator carrier. We have
introduced the phase dynamics equation as a fundamental tool in describing the

Fig. 4.9 Simulated and
calculated transconductor
thermal noise at 1 MHz offset
from the oscillator carrier.
The calculation follows
Eq. (4.35). The x-axis shows
the transconductor device
width

Fig. 4.10 Simulated and
calculated phase noise due to
the transconductor thermal
noise at 1 MHz offset from
the oscillator carrier against
transconductor device width.
The calculation is done
according to Eq. (4.38)

Fig. 4.11 Calculated
cyclostationary factor F_{cs}
against device width. The
calculation is done according
to Eq. (4.36)

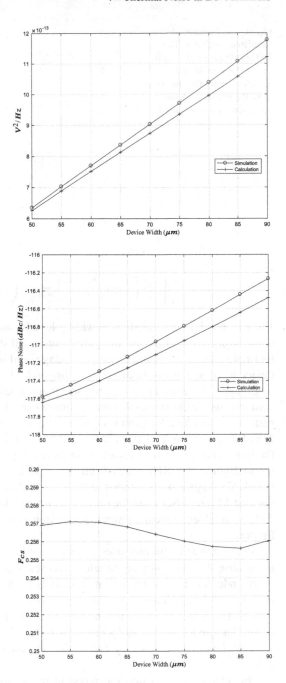

conversion of the translated noise to phase noise. We have verified that the proposed
approach gives the correct result when applied to the case of stationary tank noise
(Leeson's formula). We have subsequently applied the proposed methodology to

derive an expression for the phase noise due to the transconductor thermal noise. We have demonstrated that the predicted results agree closely with the simulation. The proposed method is not meant to replace simulators but to highlight the physical mechanisms that govern noise to phase noise conversion in oscillators. In the following chapter, we apply the phase dynamics equation to the case of the transconductor and bias/supply low-frequency noise.

Chapter 5
Low-Frequency Noise in LC Oscillators

5.1 Introduction

In the present chapter, we apply the phase dynamics equation to analyze the conversion of various types of low-frequency noise to phase noise in LC-tuned oscillators. We focus on three cases of practical interest, namely transconductor flicker noise, supply- and bias-circuitry flicker noise, and noise due to voltage-dependent capacitors. We demonstrate that the impact of such low-frequency noise sources on phase noise can be alleviated by properly designing the common-mode impedance of the LC tank at the oscillator second harmonic.

5.2 The Modified Phase Dynamics Equation

When analyzing the conversion of low-frequency noise to phase noise, it is important to know the phase relationship between the first harmonic of the oscillator carrier and the first harmonic of the transconductor current. This can be obtained by substituting (2.60)

$$v(t) \approx A_1 \cos(\omega t) - A_3 \sin(3\omega t) \tag{5.1}$$

into (2.16)

$$G(v) = i(v) = g_1 v - g_3 v^3 \tag{5.2}$$

Evaluating the 1st harmonic of the transconductor current gives

$$I_1 \approx \left(g_1 A_1 - \frac{3}{4} g_3 A_1^3 \right) \cos \omega t + \frac{9}{128} \frac{g_3^2 R A_1^5}{Q} \sin \omega t \tag{5.3}$$

In (5.3), we have explicitly indicated the dependence of I_1 on A_1 using (2.62)

$$A_3 = \frac{3g_3R}{32Q}A_1{}^3 \tag{5.4}$$

The first harmonic of the transconductor current I_1, therefore, lags the first harmonic of the tank voltage V_1 by

$$\theta_1 \approx \arctan\left(\frac{(g_1R - 1)^2}{8Q}\right) \tag{5.5}$$

In deriving (5.5), we have used the expression for A_1 in (2.61)

$$A_1 = \sqrt{\frac{4(g_1R - 1)}{3g_3R}} \tag{5.6}$$

The physical interpretation of (5.5) stems from the fact that the oscillation frequency is not the tank center frequency ω_o, but slightly lower as is shown in (2.57)

$$\omega = \left(1 - \frac{1}{16}\epsilon^2\right)\omega_o \tag{5.7}$$

where ϵ is given in (2.58)

$$\epsilon = \frac{g_1R - 1}{Q} \tag{5.8}$$

The phase of the tank admittance for small frequency deviations around ω_o can be approximated by

$$\angle Y_{tank}(\omega) \approx \arctan\left(2Q\frac{\omega - \omega_o}{\omega_o}\right) \tag{5.9}$$

Substituting (5.7) and (5.8) into (5.9) gives (5.5).

Equation (5.3) furthermore invites the introduction of the describing function approximation

$$G_D = \frac{I_1}{A_1} = \left(g_1 - \frac{3}{4}g_3A_1{}^2\right)\cos\omega t + \frac{9}{128}\frac{g_3{}^2RA_1{}^4}{Q}\sin\omega t \tag{5.10}$$

Such an approximation is valid, as the first harmonic dominates the oscillator spectrum. The incremental change ΔI_1 in I_1 due to a small disturbance ΔA_1 in A_1 can be expressed by the differential

Fig. 5.1 If a small-signal current i is injected across the tank leading the carrier by $\Delta\theta_1$, the carrier deviation ΔA_1 is in phase with the carrier A_1

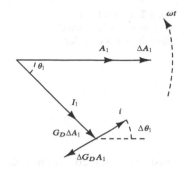

$$\Delta I_1 = G_D \Delta A_1 + \frac{\Delta G_D}{\Delta A_1} \Delta A_1 A_1 \tag{5.11}$$

Only the first component $G_D \Delta A_1$ of ΔI_1 is in phase with I_1.

Let us assume that the deviation ΔA_1 results from the injection of an external small-signal current $i = i_o \cos(\omega t + \Delta\theta_1)$ across the tank. ΔA_1 is in phase with the carrier A_1 provided that the injected current i cancels out the second component $\frac{\Delta G_D}{\Delta A_1} \Delta A_1 A_1$ of ΔI_1 in (5.11). This is depicted in Fig. 5.1.

It follows that

$$i = -\frac{\Delta G_D}{\Delta A_1} A_1 \Delta A_1 = \frac{3}{2} g_3 A_1{}^2 \Delta A_1 \cos \omega t - \frac{9}{32} \frac{g_3{}^2 R A_1{}^4}{Q} \Delta A_1 \sin \omega t \tag{5.12}$$

from which the phase $\Delta\theta_1$ by which the injected current leads the carrier is derived as

$$\Delta\theta_1 = \arctan\left(\frac{g_1 R - 1}{4Q}\right) \tag{5.13}$$

In deriving (5.13), we have used the expression for A_1 in (5.6). From the above discussion, we conclude that if a small-signal current i is injected across the tank in phase with the carrier, the resulting carrier deviation ΔA_1 lags the carrier by $\Delta\theta_1$.

With this result in mind, we modify the phase dynamics equation by introducing the term $\Delta\theta_1$ as is shown in (5.14). The geometrical interpretation is shown in Fig. 5.2, where the injected current i_{inj} that disturbs the tank is parallel to the v-axis, while the tank response lags by $\Delta\theta_1$.

$$\frac{d\phi_{tot}}{dt} = \omega_o + \frac{i_{inj}}{A_1 C} \sin(\omega_o t - \Delta\theta_1) \tag{5.14}$$

The consideration of the terms θ_1 and $\Delta\theta_1$ given in (5.5) and (5.13), respectively, is essential when analyzing low-frequency noise to phase noise conversion and can be disregarded otherwise. Both θ_1 and $\Delta\theta_1$ become negligible for large tank quality factor Q values, and when the product $g_1 R$ approaches unity. However, we

Fig. 5.2 Geometrical
interpretation of modified
phase dynamics equation

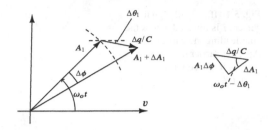

must satisfy the condition $g_1 R > 1$ in all practical oscillators to guarantee startup. Moreover, the integrated tank quality factor Q is limited in the range of 10–20 in practical tank implementations in the GHz range. These limitations result in both θ_1 and $\Delta\theta_1$ attaining finite values and therefore play a vital role in low-frequency noise to phase noise conversion [13].

5.3   Phase Noise Due to Transconductor Flicker Noise

In the following, we will assume that the drain flicker noise current density $\overline{i_{fl}^2}/df$ of a CMOS device is proportional to the drain current I_D. This is expected from the well-known expression [14]

$$\frac{\overline{v_{fl}^2}}{df} = \frac{K_f}{C_{ox}^2 \, W \, L \, f} \tag{5.15}$$

where $\overline{v_{fl}^2}$ appears in series with the gate. K_f is a mostly bias-independent quantity predominantly defined by the fabrication process. W and L are the device width and length, respectively. C_{ox} is the gate oxide capacitance. To obtain the drain flicker noise current density $\overline{i_{fl}^2}/df$, we multiply (5.15) by the square of the device transconductance. Using for the device transconductance $g_m = \sqrt{2\mu C_{ox}(W/L)I_D}$, we obtain

$$\frac{\overline{i_{fl}^2}}{df} = \frac{2\,\mu\,K_f\,I_D}{C_{ox}\,L^2\,f} = \frac{K_{fl}}{\omega}I_D \tag{5.16}$$

where $\omega = 2\pi f$, and K_{fl} is given by

$$K_{fl} = \frac{4\pi\mu K_f}{C_{ox}L^2} \tag{5.17}$$

The cyclostationary flicker noise can be expressed as

$$\frac{\overline{i_{fl_{cs}}}^2}{df} = \frac{K_{fl}}{\omega} I_D(t) \quad (5.18)$$

where $I_D(t)$ is positive. Equation (5.18) can be further written as

$$\frac{\overline{i_{fl_{cs}}}^2}{df} = \left(\frac{K_{fl} I_{Do}}{\omega}\right) \cdot \frac{I_D(t)}{I_{Do}} = \left(\frac{\overline{i_{fl_s}}^2}{df}\right) \cdot G(t) \quad (5.19)$$

I_{Do} is the device current at some instant during the oscillator period. For example, we may choose I_{Do} as the device bias current. Following (4.13), the term in the parenthesis represents stationary noise $\overline{i_{fl_s}}^2$, while the time-dependent term is the periodic function $G(t)$.

Low-frequency stationary noise components at $\Delta\omega \ll \omega_o$, where ω_o is the oscillator carrier frequency, are expressed as

$$i_{fl_s} = i_o \cdot \cos(\Delta\omega t + \psi_0) \quad (5.20)$$

where $\Delta\omega$ is close to dc. Following the convention we introduced in (4.12), $i_o^2/2$ is given by

$$i_o^2/2 = \frac{K_{fl}}{\Delta\omega} I_{Do} \quad (5.21)$$

in A^2/Hz and ψ_0 is uniformly distributed in $[0, 2\pi]$. Therefore,

$$i_{fl_{cs}} = i_{fl_s}\sqrt{G(t)} = i_{fl_s}\left(\frac{I_D(t)}{I_{Do}}\right)^{1/2} = i_{fl_s} P(t) \quad (5.22)$$

The function $P(t)$ can be expanded in a Fourier series

$$P(t) = \left(\frac{I_D(t)}{I_{Do}}\right)^{1/2} = p_0 + \sum_{l=1}^{\infty} p_l \cos(l\omega_o t + \theta_l) \quad (5.23)$$

The injected current $i_{inj} = i_{fl_s} P(t)$ into the tank is the product of (5.20) and (5.23)

$$i_{inj} = i_o \cdot \cos(\Delta\omega t + \psi_0)$$

$$\cdot \left(p_0 + \sum_{l=1}^{\infty} p_l \cos(l\omega_o t + \theta_l)\right) \quad (5.24)$$

Two further considerations allow us to simplify (5.24). First, from all the harmonics of $P(t)$, only the first upconverts noise from the baseband to the oscillator carrier frequency. Second, per (5.18), the unconverted low-frequency noise is in phase with the transconductor current. Phase θ_1 in (5.24) is therefore given by (5.5). With these simplifications, (5.24) becomes

$$i_{inj} = i_o \cos(\Delta\omega t + \psi_0) \cdot p_1 \cos(\omega_o t - \theta_1) \tag{5.25}$$

In (5.25), terms i_o and ψ_0 are defined in (5.20) and (5.21), term θ_1 is given by (5.5), and p_1 is the magnitude of the first harmonic of $P(t)$ given in (5.23). The minus sign preceding term θ_1 captures that the transconductor's first harmonic current lags the oscillator carrier.

The next step is to substitute (5.25) into the modified phase dynamics equation (5.14)

$$\frac{d\phi_{tot}}{dt} = \omega_o + \frac{i_o\,p_1}{A_1 C} \cdot$$
$$\cos(\Delta\omega t + \psi_0) \cdot \cos(\omega_o t - \theta_1) \cdot \sin(\omega_o t - \Delta\theta_1) \tag{5.26}$$

Expanding the last two terms in (5.26) and discarding the term at the second harmonic give

$$\frac{d\phi_{tot}}{dt} = \omega_o + \frac{i_o\,p_1}{2A_1 C} \cos(\Delta\omega t + \psi_0) \cdot \sin(\theta_1 - \Delta\theta_1) \tag{5.27}$$

Integrating (5.27) with zero initial conditions gives the total phase as

$$\phi_{tot} = \omega_o t + \frac{i_o\,p_1}{2A_1 C \Delta\omega} \sin(\theta_1 - \Delta\theta_1) \sin(\Delta\omega t + \psi_0) \tag{5.28}$$

The noisy oscillator carrier therefore becomes

$$A_1 \cos(\phi_{tot}) \approx A_1 \cos(\omega_o t) - \frac{i_o\,p_1}{2C \Delta\omega} \sin(\theta_1 - \Delta\theta_1) \cdot$$
$$\sin(\omega_o t) \sin(\Delta\omega t + \psi_0) \tag{5.29}$$

where the approximation is valid as the noise term is sufficiently small. Expanding the last two terms and keeping only the upper sideband give

$$A_1 \cos(\phi_{tot}) \approx A_1 \cos(\omega_o t) + \frac{i_o\,p_1}{4C \Delta\omega} \sin(\theta_1 - \Delta\theta_1) \cdot$$
$$\cos\left[(\omega_o + \Delta\omega)t + \psi_0\right] \tag{5.30}$$

The transconductor upper sideband noise power becomes

$$P_{noise} = \frac{(i_o{}^2/2)\,p_1{}^2 \sin^2(\theta_1 - \Delta\theta_1)}{32C^2(\Delta\omega)^2} \tag{5.31}$$

In deriving (5.31), we considered that two devices add noise into the tank simultaneously, while each device injects noise only on one side of the tank. The oscillator carrier power is $P_{carrier} = A_1^2/2$, so the single-sided phase noise can be expressed as

$$PN = \frac{(i_o{}^2/2)\, p_1{}^2 \, \sin^2(\theta_1 - \Delta\theta_1)}{16C^2 A_1^2 (\Delta\omega)^2} \tag{5.32}$$

Substituting for $i_o{}^2/2$, the expression given in (5.21) gives [13]

$$PN = \frac{K_{fl}\, I_{Do}\, p_1{}^2\, \sin^2(\theta_1 - \Delta\theta_1)}{16C^2 A_1^2 (\Delta\omega)^3} \tag{5.33}$$

The single-sideband phase noise due to the transconductor flicker noise reduces as the term $\theta_1 - \Delta\theta_1$ is diminished and becomes negligible as $\theta_1 - \Delta\theta_1$ goes to zero. The physical interpretation is that while transconductor flicker noise is upconverted to the vicinity of the oscillator carrier frequency due to cyclostationarity, its conversion to phase noise becomes negligible if the first harmonic of the transconductor current is aligned with the first harmonic of the oscillator carrier. This alignment is expressed in (5.33) by the term $\theta_1 - \Delta\theta_1$.

We must satisfy the condition $g_1 R > 1$ in practical oscillators to guarantee startup. Moreover, the tank quality factor Q is limited in the range of 10–20 in integrated tank implementations in the GHz range. These limitations result in both θ_1 and $\Delta\theta_1$ attaining finite values and converting transconductor flicker noise to phase noise. In a subsequent section, we will describe a method to control the term $\theta_1 - \Delta\theta_1$ to minimize the conversion of transconductor flicker noise to phase noise.

In this last part of the section, we compare (5.33) with simulation. We seek to verify that (5.33) accurately captures the physical processes that govern the conversion of transconductor flicker noise to phase noise around the oscillator carrier. We utilize the VerilogA MOS device model already discussed in Sect. 4.8. The device flicker noise is modeled by (5.16). Figure 5.3 shows the oscillator schematic used for the comparison. The tank components are chosen so that the oscillator carrier is centered at 5 GHz. The 250 Ω resistor models tank loss and is

Fig. 5.3 Schematic of the simulated oscillator. The 250 Ω resistor models tank loss and is considered noiseless. Only flicker noise from the transconductor active devices is included

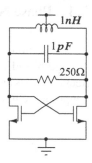

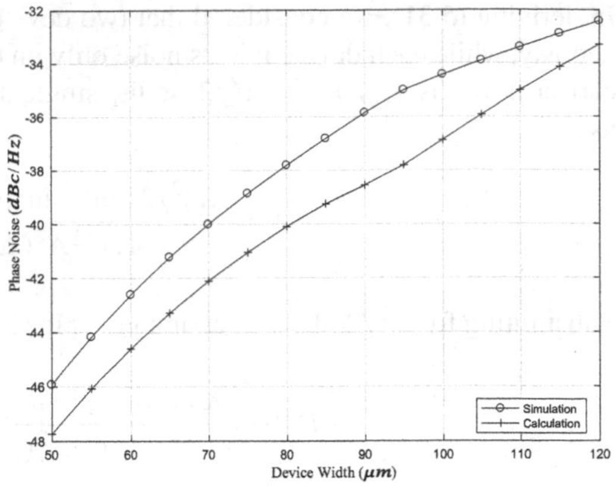

Fig. 5.4 Simulated and
calculated phase noise due to
the transconductor flicker
noise at 10 KHz offset from
the oscillator carrier against
transconductor device width.
The calculation is done
according to Eq. (5.33)

Fig. 5.5 Voltage-biased
oscillator with supply noise
$\overline{v_n}^2$

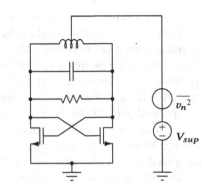

considered noiseless in the simulation. For the devices, we have used $L = 100$ nm,
$K_p = 200$ μA/V^2, $V_T = 0.35$ V, and $n = 1$. Only flicker noise from the
transconductor active devices is included. Figure 5.4 shows the phase noise due
to the transconductor flicker noise at 10 KHz offset from the oscillator carrier. The
calculation is done according to Eq. (5.33). Good agreement between simulation
and calculation is observed.

5.4 Phase Noise Due to Supply Noise

In this section, we consider the conversion of supply noise into phase noise in a
voltage-biased oscillator as shown in Fig. 5.5. As the oscillator supply is usually
heavily bypassed, we are primarily interested in the conversion of low-frequency
supply noise into phase noise.

Fig. 5.6 Low-frequency
supply noise appears in series
with the device gates

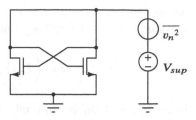

In the following, we model low-frequency supply noise as

$$\frac{\overline{v_n^2}}{df} = \frac{K_s}{f} = \frac{2\pi K_s}{\omega} \tag{5.34}$$

where K_s is the voltage noise density at 1 Hz in V^2/Hz. Low-frequency supply noise appears effectively in series with the transconductor devices' gates, as shown in Fig. 5.6. Therefore, it can be treated similarly to device flicker noise, except that both devices inject correlated noise into the tank.

Each device injects into the tank cyclostationary current noise given by

$$\frac{\overline{i_{n_{cs}}^2}}{df} = g_m^2(t) \cdot \frac{\overline{v_n^2}}{df} \tag{5.35}$$

As before, for the device transconductance, we use the expression $g_m(t) = \sqrt{2\mu C_{ox}(W/L)I_D(t)}$. Therefore, (5.35) becomes

$$\frac{\overline{i_{n_{cs}}^2}}{df} = 2\mu C_{ox}\frac{W}{L}I_D(t) \cdot \frac{2\pi K_s}{\omega} \tag{5.36}$$

This can be further written as

$$\frac{\overline{i_{n_{cs}}^2}}{df} = \left(2\mu C_{ox}\frac{W}{L}I_{Do}\frac{2\pi K_s}{\omega}\right) \cdot \frac{I_D(t)}{I_{Do}} = \left(\frac{\overline{i_{n_s}^2}}{df}\right) \cdot G(t) \tag{5.37}$$

I_{Do} is the device current at some instant during the oscillator period. For example, we may choose I_{Do} as the device bias current. Following (4.13), the term in the parenthesis represents stationary noise $\overline{i_{n_s}^2}/df$, while the time-dependent term is the periodic function $G(t)$. Low-frequency stationary noise components at $\Delta\omega \ll \omega_o$, where ω_o is the oscillator carrier frequency, are expressed as

$$i_{n_s} = i_o \cdot \cos(\Delta\omega t + \psi_0) \tag{5.38}$$

where only noise close to dc is accounted for. Following the convention we introduced in (4.12), $i_o{}^2/2$ is given by

$$i_o{}^2/2 = \left(2\mu C_{ox}\frac{W}{L}I_{Do}\right)\frac{2\pi K_s}{\Delta\omega} = g_{m_o}^2\frac{2\pi K_s}{\Delta\omega} \qquad (5.39)$$

in A^2/Hz and ψ_0 is uniformly distributed in $[0, 2\pi]$. Term g_{m_o} is the device transconductance at current I_{Do}. Therefore,

$$i_{n_{cs}} = i_{n_s}\sqrt{G(t)} = i_{n_s}\left(\frac{I_D(t)}{I_{Do}}\right)^{1/2} = i_{n_s}P(t) \qquad (5.40)$$

Following the exact analysis detailed in Eqs. (5.23) to (5.30), we obtain the noisy oscillator carrier as

$$A_1\cos(\phi_{tot}) \approx A_1\cos(\omega_o t) + 2\frac{i_o\, p_1}{4C\Delta\omega}\sin(\theta_1 - \Delta\theta_1)\cdot$$
$$\cos((\omega_o + \Delta\omega)t + \psi_0) \qquad (5.41)$$

where the factor of two in (5.41) in comparison to (5.30) accounts for the fact that the noise injected into the tank from the two devices is correlated. The upper sideband noise power becomes

$$P_{noise} = \frac{(i_o{}^2/2)\, p_1{}^2\, \sin^2(\theta_1 - \Delta\theta_1)}{16C^2(\Delta\omega)^2} \qquad (5.42)$$

In deriving (5.42), we considered that each device injects noise only on one side of the tank. The oscillator carrier power is $P_{carrier} = A_1^2/2$, and the single-sided phase noise can be expressed as

$$PN = \frac{(i_o{}^2/2)\, p_1{}^2\, \sin^2(\theta_1 - \Delta\theta_1)}{8C^2 A_1^2(\Delta\omega)^2} \qquad (5.43)$$

Substituting for $i_o{}^2/2$ the expression given in (5.39) gives

$$PN = \frac{2\pi\, g_{m_o}^2\, K_s\, p_1{}^2\, \sin^2(\theta_1 - \Delta\theta_1)}{8C^2 A_1^2(\Delta\omega)^3} \qquad (5.44)$$

Due to the low-frequency supply noise, the single-sideband phase noise reduces as the term $\theta_1 - \Delta\theta_1$ is reduced. Similarly to the case of device flicker noise, low-frequency supply noise is upconverted to the vicinity of the oscillator carrier frequency due to cyclostationarity. Its conversion to phase noise becomes negligible if the first harmonic of the transconductor current is aligned with the first harmonic of the oscillator carrier. This alignment is expressed in (5.44) by the term $\theta_1 - \Delta\theta_1$.

Fig. 5.7 Simulated and
calculated phase noise due to
low-frequency supply noise at
10 KHz offset from the
oscillator carrier against
transconductor device width.
The calculation is done
according to Eq. (5.44)

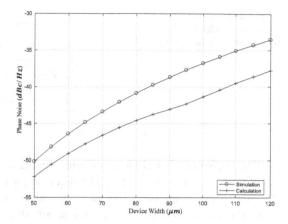

As discussed, practical limitations result in both θ_1 and $\Delta\theta_1$ attaining finite values and converting low-frequency supply noise into phase noise.

Figure 5.7 shows the simulated and calculated phase noise due to low-frequency supply noise at 10 KHz offset from the oscillator carrier against transconductor device width for the oscillator schematic shown in Fig. 5.5. The tank components (1 pF, 1 nH) are chosen to center the oscillator carrier at 5 GHz. A 250 Ω resistor models tank loss and is considered noiseless in the simulation. We have utilized the VerilogA MOS device model discussed in Sect. 4.8, with $L = 100$ nm, $K_p = 200\,\mu\text{A/V}^2$, $V_T = 0.35$ V, and $n = 1$. Only supply noise is included and is modeled according to (5.34). Phase noise calculation is done according to Eq. (5.44). A good agreement between simulation and analysis is observed.

5.5 Common-Mode Effects

In this section, we investigate the effect of the common-mode transconductance in combination with the tank common-mode impedance on the conversion of transconductor flicker noise and low-frequency supply noise into phase noise [13]. We demonstrate that such conversions can be minimized by designing the tank common-mode impedance at the second harmonic. As discussed in the previous sections, low-frequency noise emanating from the transconductor devices or the power supply upconverts to the vicinity of the oscillator carrier frequency due to cyclostationarity. Its conversion to phase noise is governed by the phase dynamics equation, which we used to derive Eqs. (5.33) and (5.44). These two equations suggest that aligning the first harmonic of the transconductor current with the first harmonic of the oscillator carrier reduces the conversion of transconductor flicker noise and low-frequency supply noise into phase noise, respectively. This alignment is expressed mathematically by the term $\theta_1 - \Delta\theta_1$ in (5.33) and (5.44).

Fig. 5.8 LC self-sustained
oscillator model with
common-mode impedance

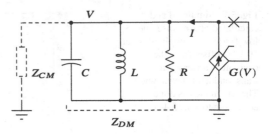

The impact of the oscillator differential operation on θ_1 and $\Delta\theta_1$ has already been
shown in Eqs. (5.5) and (5.13). The requirement to satisfy the condition $g_1R > 1$
to guarantee startup, as well as the finite values of tank quality factors achieved in
practical integrated oscillators, results in both θ_1 and $\Delta\theta_1$ attaining non-zero values
and subsequently in the conversion of low-frequency noise into phase noise.

To extend Eqs. (5.5) and (5.13) to account for the oscillator common-mode
operation, we work with the Van der Pol self-sustained oscillator model shown in
Fig. 5.8. The tank differential impedance Z_{DM} consists of the LC tank and the tank
loss resistor R. Z_{CM} models the tank common-mode impedance. $G(V)$ is modeled
as

$$G(V) = g_1V - g_3V^3 + g_2V^2 \tag{5.45}$$

Term g_2 approximates the common-mode behavior of the transconductor. In the
following, we extend the method presented in Sect. 2.6 to account for common-
mode effects [13]. We start the analysis by opening the positive feedback loop at
the control terminal of the nonlinear current source $G(V)$ and applying a voltage
$A_1\cos(\omega t)$. The current I flowing out of $G(V)$ into the tank becomes

$$I = \left(A_1g_1 - \frac{3}{4}A_1^3g_3\right)\cos(\omega t) - \frac{1}{4}A_1^3g_3\cos(3\omega t) + \frac{1}{2}A_1^2g_2\cos(2\omega t) \tag{5.46}$$

The odd harmonics flow into the tank differential impedance Z_{DM}, while the second
harmonic flows into the tank common-mode impedance Z_{CM}. The tank voltage
therefore becomes

$$V = \left(A_1g_1 - \frac{3}{4}A_1^3g_3\right)R\cos(\omega t)$$

$$- \frac{3}{32Q}A_1^3g_3R\sin(3\omega t)$$

$$+ \frac{1}{2}A_1^2g_2Z_{CM}|_{2\omega}\cos(2\omega t) \tag{5.47}$$

where we used for the tank quality factor Q at the oscillator's first harmonic and for
the tank impedance at the oscillator's third harmonic

$$Q = \omega RC \qquad (5.48)$$

$$Z_{LC}|_{3\omega} = -j\frac{3}{8\omega C} \qquad (5.49)$$

In (5.47), the term $Z_{CM}|_{2\omega}$ denotes the tank common-mode impedance at the oscillator second harmonic. In most practical cases, the real part of the common-mode impedance at the second harmonic is small, which allows us to express (5.47) as

$$V = \left(A_1 g_1 - \frac{3}{4}A_1{}^3 g_3\right) R \cos(\omega t)$$

$$-\frac{3}{32Q}A_1{}^3 g_3 R \sin(3\omega t)$$

$$-\frac{1}{2}A_1{}^2 g_2 Z_{Cmi} \sin(2\omega t) \qquad (5.50)$$

where Z_{Cmi} is the imaginary part of the common-mode impedance at the second harmonic. The transconductor first harmonic current I_1 can be estimated by substituting (5.50) into (5.45) giving

$$I_1 \approx \left(g_1 A_1 - \frac{3}{4}g_3 A_1{}^3\right)\cos(\omega t) + \frac{9}{128}\frac{g_3{}^2 R A_1{}^5}{Q}\sin(\omega t)$$

$$-\frac{1}{2}g_2{}^2 A_1{}^3 Z_{CMi}\sin(\omega t) \qquad (5.51)$$

In (5.51), we have explicitly indicated the dependence of I_1 on A_1 using (5.4). Equation (5.51) is the same as (5.3) with an additional quadrature term that accounts for common-mode effects. The first harmonic of the transconductor current I_1, therefore, lags the first harmonic of the tank voltage V_1 by

$$\theta_1 \approx \arctan\left(\frac{(g_1 R - 1)^2}{8Q} - \frac{2g_2{}^2 Z_{CMi}}{3g_3}(g_1 R - 1)\right) \qquad (5.52)$$

In deriving (5.52), we have used the expression for A_1 in (5.6). Equation (5.52) is, of course, the same as (5.5) except for the additional term that accounts for common-mode effects.

Similarly to (5.3), Eq. (5.51) invites the introduction of the describing function approximation

$$G_D = \frac{I_1}{A_1} = \left(g_1 - \frac{3}{4}g_3 A_1{}^2\right)\cos\omega t + \frac{9}{128}\frac{g_3{}^2 R A_1{}^4}{Q}\sin\omega t$$

$$-\frac{1}{2}g_2{}^2 A_1{}^2 Z_{CMi}\sin(\omega t) \qquad (5.53)$$

The incremental change ΔI_1 in I_1 due to a small disturbance ΔA_1 in A_1 can be expressed by the differential

$$\Delta I_1 = G_D \Delta A_1 + \frac{\Delta G_D}{\Delta A_1} \Delta A_1 A_1 \tag{5.54}$$

Only the first component $G_D \Delta A_1$ of ΔI_1 is in phase with I_1. Let us assume that the deviation ΔA_1 results from the injection of an external small-signal current $i = i_o \cos(\omega t + \Delta \theta_1)$ across the tank. ΔA_1 is in phase with the carrier A_1 provided that the injected current i cancels out the second component $\frac{\Delta G_D}{\Delta A_1} \Delta A_1 A_1$ of ΔI_1 in (5.54). It follows that

$$i = -\frac{\Delta G_D}{\Delta A_1} A_1 \Delta A_1$$

$$= \frac{3}{2} g_3 A_1^2 \Delta A_1 \cos \omega t - \frac{9}{32} \frac{g_3^2 R A_1^4}{Q} \Delta A_1 \sin \omega t$$

$$+ g_2^2 Z_{CMi} A_1^2 \Delta A_1 \sin \omega t \tag{5.55}$$

from which the phase $\Delta \theta_1$ by which the injected current leads the carrier is derived as

$$\Delta \theta_1 = \arctan \left(\frac{g_1 R - 1}{4Q} - \frac{2 g_2^2 Z_{CMi}}{3 g_3} \right) \tag{5.56}$$

In deriving (5.56), we have used the expression for A_1 in (5.6). From the above discussion, we conclude that if a small-signal current i is injected across the tank in phase with the carrier, the resulting carrier deviation ΔA_1 lags the carrier by $\Delta \theta_1$. Equation (5.56) is, of course, the same as (5.13) except for the additional term that accounts for common-mode effects.

In the presence of common-mode effects, the term $\theta_1 - \Delta \theta_1$ is given by summing (5.52) and (5.56). As $\theta_1 - \Delta \theta_1$ approaches zero, the transconductor flicker noise and the low-frequency supply/bias noise conversion into phase noise are nulled. This condition is realized by properly adjusting the imaginary part of the tank's common-mode impedance at the second harmonic. In practice, placing the common-mode resonance close to the oscillator's second harmonic allows us to realize the desired result.

We have simulated the oscillator shown in Fig. 5.9 to verify the above analysis. Two tightly coupled filaments of self-inductance L_f constitute the tank differential inductance. L_{CM} is the common-mode inductance. C_{DM} and C_{CM} are the tank differential- and common-mode capacitances, respectively. Resistor R models tank loss. Devices M_1 and M_2 are modeled using the VerilogA MOS device model already discussed in previous sections. Their flicker noise is modeled according to (5.16). Voltage noise source $\overline{v_n^2}$ captures supply/bias noise according to (5.34). The oscillator is tuned at 5 GHz.

Fig. 5.9 Oscillator schematic
capturing the tank
common-mode behavior

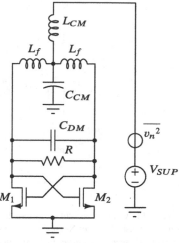

Fig. 5.10 Continuous curve:
Phase noise due to device
flicker noise and
low-frequency supply noise at
10 KHz offset from the
oscillator carrier. Dashed
curve: $\theta_1 - \Delta\theta_1$ in deg.
X-axis: Tank common-mode
inductance L_{CM} in (pH).
Phase noise nulls as the term
$\theta_1 - \Delta\theta_1$ crosses zero

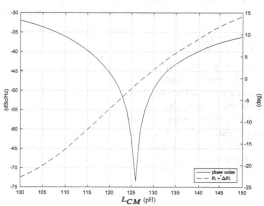

Figure 5.10 shows the phase noise due to device flicker noise and low-frequency
supply noise at 10 KHz offset from the oscillator carrier. The term $\theta_1 - \Delta\theta_1$ is also
plotted. Phase noise nulls as $\theta_1 - \Delta\theta_1 \approx 0$ as expected from the analysis. Figure 5.11
depicts the device flicker noise and supply noise contributions at 10 KHz offset from
the oscillator carrier. Both null as the term $\theta_1 - \Delta\theta_1$ crosses zero in agreement with
the analysis.

5.6   Phase Noise Due to Varactors

This last section discusses the upconversion of low-frequency noise due to varactors.
Without loss of generality, we concentrate on the oscillator shown in Fig. 5.12,
where capacitors C_C couple the varactors C_V to the tank, and assume that $C_C \gg$
C_V. The large bias resistors R_B facilitate the dc biasing of the varactors.

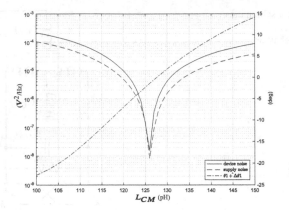

Fig. 5.11 Continuous curve: Noise due to device flicker noise at 10 KHz offset from the oscillator carrier. Dashed curve: Noise due to low-frequency supply noise at 10 KHz offset from the oscillator carrier. Dashed–dot curve: $\theta_1 - \Delta\theta_1$ in deg. X-axis: Tank common-mode inductance L_{CM} in (pH). The device flicker noise and the low-frequency supply noise null as the term $\theta_1 - \Delta\theta_1$ crosses zero

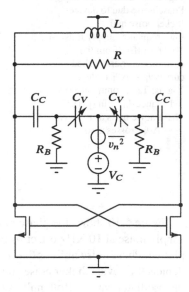

Fig. 5.12 Oscillator schematic showing the varactors C_V in the tank. Low-frequency noise emanating from the supply or from the circuitry that controls the common terminal of the varactors is modeled as $\overline{v_n^2}$ and is placed in series with the varactor control voltage V_C

For the varactors, we assume a $C - V$ relationship of the form

$$C_V = C_0 + C_1 \, V_V + C_3 \, V_V{}^3 \tag{5.57}$$

where V_V is the voltage across the varactor. The third-order term is introduced to capture the flattening of the $C - V$ characteristic for large V_V. The instantaneous voltage across each varactor is half the tank differential voltage given by (2.60)

$$v(t) \approx A_1 \cos(\omega t) - A_3 \sin(3\omega t) \tag{5.58}$$

minus the dc voltage at the common control terminal V_C. Therefore,

$$V_V \approx \frac{A_1}{2} \cos(\omega t) - \frac{A_3}{2} \sin(3\omega t) - V_C \tag{5.59}$$

Substituting (5.59) into (5.57) and averaging over the oscillator period give the average varactor capacitance as

$$C_{V_{av}} = C_0 - C_1 V_C - C_3 V_C^3 - \frac{3}{8}(A_1^2 + A_3^2)C_3 V_C \tag{5.60}$$

Low-frequency noise emanating from the supply or from the circuitry that controls the common terminal of the varactors is modeled as $\overline{v_n^2}$ and is placed in series with the control voltage V_C in Fig. 5.12. $\overline{v_n^2}$ upconverts into the oscillator carrier frequency due to frequency modulation. The noisy carrier $V_c(t)$ can be expressed as

$$V_c(t) = A_1 \cos\left(\omega t + 2\pi K_m \int_{-\infty}^{t} v_n(\tau)d\tau\right) \tag{5.61}$$

$v_n(t)$ denotes the low-frequency noise appearing in series with the varactor control voltage V_C, and K_m is the modulation index in Hz/V. To estimate K_m, we need to calculate the shift in the LC tank center frequency $f_o = 1/2\pi\sqrt{LC_0}$ due to a small change ΔC in the tank capacitance. This is given by

$$f_o + \Delta f = \frac{1}{2\pi\sqrt{L(C_0 - \Delta C)}} \approx f_o \cdot \left(1 + \frac{\Delta C}{2C_0}\right) \tag{5.62}$$

(5.62) gives for Δf

$$\Delta f \approx f_o \frac{\Delta C}{2C_0} \tag{5.63}$$

The modulation index K_m can be expressed as

$$K_m = \frac{\Delta f}{\Delta V_C} = \frac{\Delta f}{\Delta C} \cdot \frac{\Delta C}{\Delta V_C} = \frac{f_o}{2C_0} \cdot \frac{\Delta C_{V_{av}}}{\Delta V_C} \tag{5.64}$$

Differentiating Eq. (5.60) with respect to the varactor control voltage V_C and substituting into (5.64) give

$$K_m = -\frac{f_o}{2C_0} \cdot \left(C_1 + \frac{3}{8}C_3(A_1^2 + A_3^2) + 3C_3 V_C^2\right) \tag{5.65}$$

We model the low-frequency noise spectral density entering the varactor control terminal by

$$\frac{\overline{v_n}^2}{df} = \frac{K_v}{f} = \frac{2\pi K_v}{\omega} \qquad (5.66)$$

K_v is the noise spectral density at 1 Hz in V^2/Hz. Following the convention we introduced in (4.12), noise components at frequency $\Delta\omega$ are written as

$$v_n = v_o \cdot \cos(\Delta\omega t + \psi) \qquad (5.67)$$

where

$$\frac{v_o^2}{2} = \frac{2\pi K_v}{\Delta\omega} \qquad (5.68)$$

is in V^2/Hz and ψ is uniformly distributed in $[0, 2\pi]$. Substituting (5.67) into (5.61) and assuming that the noise term is adequately small give for the noisy carrier

$$V_c(t) \approx A_1 \cos(\omega t) + 2\frac{2\pi K_m A_1 v_o}{2\Delta\omega} \cos\left((\omega \pm \Delta\omega)t \pm \psi)\right) \qquad (5.69)$$

The factor of two is included to account for the fact that the two varactors inject correlated noise into the tank. The single-sideband phase noise can thus be expressed as

$$PN = \frac{(2\pi)^2 K_m^2 (v_o^2/2)}{2(\Delta\omega)^2} \qquad (5.70)$$

In deriving (5.70), we considered that each varactor injects noise only on one side of the tank. Substituting (5.65) and (5.68) into (5.70) gives

$$PN = \left(\frac{f_o}{2C_0}\right)^2 \cdot \left(C_1 + \frac{3}{8}C_3(A_1^2 + A_3^2) + 3C_3 V_C^2\right)^2 \cdot \frac{K_v}{2(\Delta f)^3} \qquad (5.71)$$

In this last part of the section, we compare (5.71) with the simulation of the oscillator depicted in Fig. 5.12. We utilize the VerilogA MOS device model already discussed in previous sections. The tank components are chosen so that the oscillator carrier is centered at 5 GHz. For the varactor, we have used $C_0 = 2$ pF, $C_1 = 40$ fF, and $C_3 = -15$ fF. Resistor $R = 250\ \Omega$ models tank loss and is considered noiseless in the simulation. The large biasing resistors R_B are also noiseless. Figure 5.13 compares the simulated and estimated phase noise at 10 KHz offset from the oscillator carrier due to low-frequency noise at the varactor control port. This is done for different values of the varactor control voltage shown on the x-axis. A very good agreement between simulation and calculation is observed.

In this and the previous chapters, we have demonstrated the phase dynamics equation's central role in accurately describing phase noise in oscillators. Our analysis captures the fundamental mechanisms behind noise to phase noise conversion.

Fig. 5.13 Simulated and
calculated phase noise at 10
KHz offset from the oscillator
carrier due to low-frequency
noise appearing at the
varactor port against the
varactor control voltage. The
calculation is done according
to Eq. (5.71)

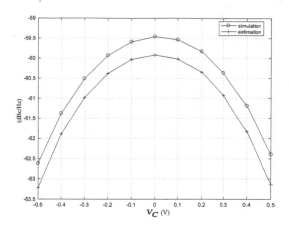

As pointed out, it does not replace simulators but highlights the physical phenomena
and offers design insight. The description of an oscillator as a rotating phase point is
powerful. In the next chapter, we capitalize on this by applying the phase dynamics
equation to describe oscillator entrainment and pulling.

Chapter 6
LC Oscillator Entrainment and Pulling

6.1 Introduction

Injecting a current across a RLC tank generates a spur at the injection frequency. Injecting a current across the tank of a self-sustained oscillator is more complicated. If the injection happens close to the oscillator free-running frequency, it results in injection locking. Therefore, the oscillator carrier is entrained and is frequency- and phase-locked to the injected signal. The injected current cannot entrain the oscillator if the injection happens further from the oscillator's free-running frequency. The oscillator's motion is still disturbed, resulting in the generation of multiple spurs. Furthermore, if the injected current is at the same frequency as the oscillator carrier with an additional phase shift, the oscillation frequency shifts away from the value defined by the oscillator tank components; this is known as oscillator pulling. In the first part of the chapter, we delve into these effects. We concentrate on the case where the injection results from the magnetic coupling. In the second part of the chapter, we focus on magnetic coupling and look at ways to reduce it.

6.2 Magnetic Coupling on *RLC* Tank

Figure 6.1 depicts a RLC tank aggressed by current i flowing in a nearby inductor L_2. The magnetic coupling factor k between L_2 and the tank inductor L_1 is given by $k = M/\sqrt{L_1 L_2}$, where M is the mutual inductance.

The arrangement in Fig. 6.1 can be modeled as is shown in Fig. 6.2. The voltage developed across the RLC tank due to the current i flowing in L_2 can be expressed as

$$v = \frac{M}{L_1} i \cdot Z_{RLC} \tag{6.1}$$

K. Manetakis, *Topics in LC Oscillators*,
https://doi.org/10.1007/978-3-031-31086-7_6

Fig. 6.1 Magnetic coupling
between L_2 and RLC tank

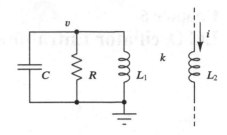

Fig. 6.2 Equivalent model of
magnetic coupling between
L_2 and RLC tank

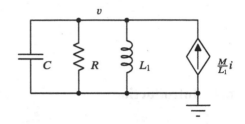

where Z_{RLC} is the impedance of the tank given in (1.42)

$$Z_{RLC}(s) = \frac{s/C}{s^2 + \beta s + \omega_o^2} \tag{6.2}$$

where $\beta = 1/RC$ and $\omega_o{}^2 = 1/LC$. We therefore have

$$v = \frac{M}{L_1} i \cdot \frac{s/C}{s^2 + \beta s + \omega_o^2} \approx \frac{M}{L_1} i \cdot R \tag{6.3}$$

where the approximation is valid close to the tank resonance frequency. For
example, the RLC tank may be part of a low-noise amplifier. The current i flowing
in the inductor L_2 models the high-frequency harmonics of a clock signal that
appear within the amplifier's frequency range of operation. The finite magnetic
coupling between L_2 and the amplifier RLC tank generates spurs described by
Eq. (6.3). In practice, the clock harmonics may be a few μA, and the mutual
inductance a few pH. Depending on the sensitivity level of the receiver, even such
minute effects may be detrimental, and the coupling is therefore unwanted. Later
sections discuss measures often taken to minimize such magnetic aggression.

6.3   Magnetic Coupling on Self-Sustained Oscillator

Figure 6.3 depicts a self-sustained oscillator aggressed by current i flowing in a
nearby inductor L_2. An equivalent model is shown in Fig. 6.4, where the aggressing

Fig. 6.3 Magnetic coupling between L_2 and tank of self-sustained oscillator

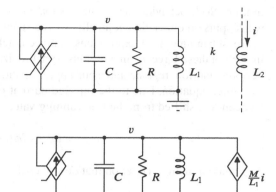

Fig. 6.4 Equivalent model of magnetic coupling between L_2 and tank of self-sustained oscillator

current is injected across the oscillator tank. The carrier of the undisturbed oscillator is

$$v = A_1 \cdot \cos(\omega_o t), \tag{6.4}$$

where ω_o is the oscillator free-running frequency and A_1 is the oscillation amplitude, both determined by its internal parameters.

The current flowing in inductor L_2 is assumed to be $i = i_o \cdot \cos(\omega_1 t)$. The current injected across the oscillator tank in Fig. 6.4 becomes

$$i_{inj} = \frac{M}{L_1} i_o \cdot \cos(\omega_1 t) = i_{inj_o} \cdot \cos[(\omega_o + \Delta\omega)t] \tag{6.5}$$

where the detuning $\Delta\omega = \omega_1 - \omega_o$ quantifies the difference in frequency between the free-running oscillator and the injected current. In the following, we limit ourselves to cases where the aggressing current constitutes a weak disturbance. By this, we mean that it only affects the oscillator phase but not the oscillator amplitude. The oscillator retains control of its amplitude, while its phase is disturbed by the injected current.

As discussed in Chap. 2, the oscillator motion can be portrayed as a phase point moving along the oscillator's limit cycle. The external current disturbs the oscillator's motion [4]. For adequately small detuning values, the external current can entrain the oscillator. The oscillator and the external current are then synchronized. We say that the oscillator is injection-locked by the external current. For larger detuning values, the external current cannot entrain the oscillator but still affects its motion. The oscillator then undergoes a complex motion. Its phase point is alternately accelerated and decelerated as it moves along the limit cycle, exhibiting what is known as a quasi-periodic motion [4]. The resulting spectrum is multiple spurs around the oscillator's free-running frequency. As the detuning is further increased, the oscillator motion is very weakly disturbed from its free motion. Such behavior can be approximated by the appearance of well-separated, weak spurs on either side of an almost free-running oscillator carrier.

To obtain analytical expressions that accurately describe the oscillator behavior in each of these three distinct regions: synchronization or entrainment region, quasi-periodic motion region, and spur-approximation region, we consider the phase dynamics equation. Due to the injected current i_{inj} across the tank, the oscillation frequency is shifted from the free-running value ω_o and becomes

$$\omega_{osc} = \omega_o + d\phi(t)/dt \qquad (6.6)$$

The phase dynamics equation that describes the evolution of $\phi(t)$ is written as

$$\frac{d\phi(t)}{dt} = \frac{1}{A_1 C} \cdot \left[i_{inj_o} \cdot cos[(\omega_o + \Delta\omega)t] \cdot sin[\omega_o t + \phi(t)]\right] \qquad (6.7)$$

The term in the brackets is the injected current. To account for the frequency shift, we have included $\phi(t)$ inside the argument of the sin term together with $\omega_o t$. C is the oscillator tank capacitance, while A_1 is the carrier amplitude. Expanding (6.7) and discarding the term around $2\omega_o$ give

$$\frac{d\phi(t)}{dt} = \frac{i_{inj_o}}{2A_1 C} \cdot sin[\phi(t) - \Delta\omega \cdot t] \qquad (6.8)$$

Due to the sin term on the right-hand side of (6.8), we expect that the frequency shift $d\phi(t)/dt$ is bounded by

$$\left| \frac{d\phi(t)}{dt} \right| \le \frac{i_{inj_o}}{2A_1 C} = \alpha \qquad (6.9)$$

Solving (6.8) gives two distinct solutions depending on the value of the detuning $\Delta\omega$ relative to α [1]. For $\Delta\omega < \alpha$, we obtain

$$\phi(t) = \Delta\omega \cdot t - 2 \arctan\left[\frac{-\alpha - \sqrt{\alpha^2 - \Delta\omega^2} \cdot \tanh\left[\frac{1}{2}\sqrt{\alpha^2 - \Delta\omega^2} \cdot t - arctanh(\frac{\alpha}{\sqrt{\alpha^2 - \Delta\omega^2}}) \right]}{\Delta\omega} \right]$$

$$(6.10)$$

while for $\Delta\omega > \alpha$, we get

$$\phi(t) = \Delta\omega \cdot t - 2 \arctan\left[\frac{-\alpha + \sqrt{\Delta\omega^2 - \alpha^2} \cdot \tan\left[\frac{1}{2}\sqrt{\Delta\omega^2 - \alpha^2} \cdot t - arctan(\frac{-\alpha}{\sqrt{\Delta\omega^2 - \alpha^2}}) \right]}{\Delta\omega} \right]$$

$$(6.11)$$

---

[1] The solution of (6.8) is outlined in Appendix B.

Equation (6.10) is valid for detuning $\Delta\omega$ smaller than the critical value α and corresponds to the entrainment region. Equation (6.11) is valid when the detuning $\Delta\omega$ exceeds α and applies to the quasi-periodic motion and the spur-approximation regions. The critical value of the detuning α given in (6.9) defines the locking range. It is proportional to the amplitude of the injected current i_{inj_o} and shrinks as the oscillator amplitude A_1 and tank capacitance C increase. In the following two sections, we will delve into each of the solutions given by Eqs. (6.10) and (6.11), compare them with simulations, and discuss their practical implications.

6.4   Entrainment of Self-Sustained Oscillator

In this region, the detuning $\Delta\omega$ is smaller than the critical value α, and the oscillator phase is governed by Eq. (6.10). Figure 6.5 shows an example of the time evolution of the oscillator phase $\phi(t)$ for different detuning values. The oscillator amplitude is $A_1 = 1$ V, the tank capacitance is $C = 1$ pF, and the amplitude of the injected current across the tank is $i_{inj_o} = 25$ uA. The critical value of the detuning given by (6.9) is, therefore, $\alpha/(2\pi) = 2$ MHz. After an initial transient, the oscillator phase $\phi(t)$ increases linearly with time. From Eq. (6.6), the slope $d\phi(t)/dt$ equals the oscillator frequency shift. Therefore, Fig. 6.5 indicates the constant frequency shift. Furthermore, Fig. 6.6 shows that it is equal to the detuning, confirming that the injected current entrains the oscillator.

Figure 6.7 shows the simulated and calculated spectra of the oscillator depicted in Fig. 6.4. The continuous curve is the simulated spectrum of the free-running oscillator. The free-running frequency is 5028.4 MHz. The dash–dot curve depicts the simulated oscillator spectrum when an external current is injected across the tank. The frequency of the injected current is 5029 MHz, while $A_1 = 1$ V, $C = 1$ pF, and $i_{inj_o} = 25$ uA. The critical value of the detuning is $\alpha/(2\pi) = 2$ MHz. The detuning is, therefore, smaller than 2 MHz, resulting in injection locking of

Fig. 6.5 Entrained oscillator phase $\phi(t)$ for different values of the detuning. $A_1 = 1$ V, $C = 1$ pF, $i_{inj_o} = 25$ uA. The critical value of the detuning is $\alpha/(2\pi) = 2$ MHz

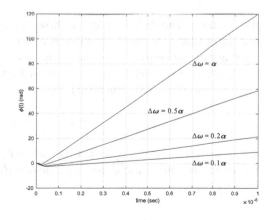

Fig. 6.6 Oscillator frequency
shift as a function of the
detuning. $A_1 = 1$ V,
$C = 1$ pF, $i_{inj_o} = 25$ uA. The
critical value of the detuning
is $\alpha/(2\pi) = 2$ MHz. As the
detuning is smaller than 2
MHz, the oscillator is
entrained

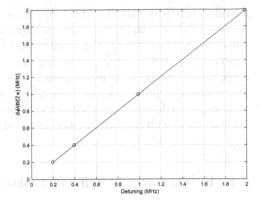

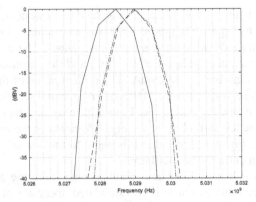

Fig. 6.7 Continuous curve: Free-running oscillator spectrum from transient simulation. Dash–dot curve: entrained oscillator spectrum from transient simulation. Dash–dash curve: entrained oscillator spectrum, where the oscillator phase is calculated by Equations (6.10) and (6.6). $A_1 = 1$ V, $C = 1$ pF, $i_{inj_o} = 25$ uA. The critical value of the detuning is $\alpha/(2\pi) = 2$ MHz. The oscillator's free-running frequency is 5028.4 MHz. The frequency of the injected current is 5029 MHz. The detuning is, therefore, smaller than 2 MHz, resulting in injection locking of the oscillator

the oscillator. The dash–dash curve shows the calculated spectrum of the entrained oscillator, where the oscillator phase is obtained from Eqs. (6.10) and (6.6). A very good agreement between transient simulation and analysis is observed.

Equation (6.10) and Fig. 6.5 suggest that after the initial transient has decayed, the oscillator phase $\phi(t)$ can be expressed as

$$\phi(t) = \Delta\omega \cdot t + \psi \tag{6.12}$$

where ψ is a constant angle that depends only on the initial transient. Substituting (6.12) into the phase dynamics equation (6.8) and using the expression for α in (6.9) give

Fig. 6.8 Continuous curve: Injected current across the oscillator tank. Dashed curve: entrained oscillator tank voltage. The oscillator carrier is frequency- and phase-locked to the injected current

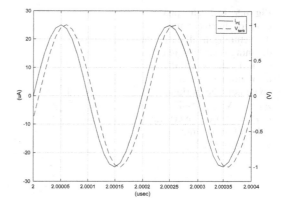

$$\psi = arcsin\left(\frac{\Delta\omega}{\alpha}\right) \pm n\pi \qquad (6.13)$$

where n is an integer. Equation (6.13) shows that the entrained oscillator carrier locks to the phase of the injected signal. This is clearly shown in Fig. 6.8. For the specific case of zero detuning $\Delta\omega = 0$, phase locking happens with an uncertainty of π, meaning that the injection and response are either in-phase or anti-phase.

6.5 Quasi-periodic Motion Region

In this region, the detuning $\Delta\omega$ is larger than the critical value α, and the oscillator phase is governed by Eq. (6.11). Figure 6.9 shows an example of the time evolution of the oscillator phase $\phi(t)$ for different detuning values. The oscillator amplitude is $A_1 = 1$ V, the tank capacitance is $C = 1$ pF, and the amplitude of the injected current across the tank is $i_{inj_o} = 25$ uA. The critical value of the detuning given by (6.9) is $\alpha/(2\pi) = 2$ MHz. After an initial transient, the oscillator phase $\phi(t)$ increases with time in a wiggly fashion, indicating a complex motion of the oscillator phase point along its limit cycle. The external current cannot entrain the oscillator but still affects it, resulting in it undergoing a quasi-periodic motion [4].

The average slope $d\phi(t)/dt$ is smaller than the critical value α. The oscillator cannot quite follow the injected signal, and as the detuning increases, the oscillator's response to the injection signal diminishes. For adequately large detuning values, the oscillator carrier is only weakly disturbed by the injection signal, as shown in Fig. 6.10.

Fig. 6.9 Oscillator phase $\phi(t)$ for different values of the detuning. In all cases, the detuning is larger than the critical value α, and the oscillator undergoes quasi-periodic motion. $A_1 = 1$ V, $C = 1$ pF, $i_{inj_o} = 25$ uA. The critical value of the detuning is $\alpha/(2\pi) = 2$ MHz

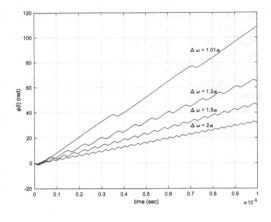

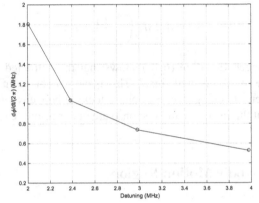

Fig. 6.10 Oscillator average slope $d\phi(t)/dt$ for different values of the detuning. In all cases, the detuning is larger than the critical value α, and the oscillator undergoes quasi-periodic motion. $A_1 = 1$ V, $C = 1$ pF, $i_{inj_o} = 25$ uA. The critical value of the detuning is $\alpha/(2\pi) = 2$ MHz. The oscillator cannot quite follow the injected signal, and as the detuning increases, the oscillator's response to the injection signal diminishes

Figure 6.11 shows the simulated and calculated spectra of the oscillator depicted in Fig. 6.4 undergoing quasi-periodic motion. $A_1 = 1$ V, $C = 1$ pF, $i_{inj_o} = 25$ uA. The critical value of the detuning is $\alpha/(2\pi) = 2$ MHz. The continuous curve depicts the simulated spectrum. The dashed curve shows the calculated oscillator spectrum, where the phase is obtained from Eqs. (6.11) and (6.6). The oscillator free-running frequency is 5028.4 MHz, while the frequency of the injected current is 5031 MHz. The detuning is, therefore, larger than 2 MHz, and the injected current fails to entrain the oscillator. A very good agreement between transient simulation and analysis is observed.

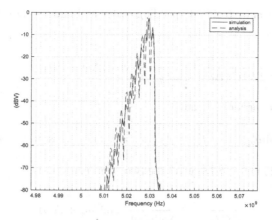

Fig. 6.11 Continuous curve: oscillator spectrum undergoing quasi-periodic motion (transient simulation). Dashed curve: oscillator spectrum undergoing quasi-periodic motion, where the oscillator phase is calculated by Eqs. (6.11) and (6.6). In this example, $A_1 = 1$ V, $C = 1$ pF, $i_{inj_o} = 25$ uA. The critical value of the detuning is $\alpha/(2\pi) = 2$ MHz. The frequency of the injected current is 5031 MHz. The detuning is, therefore, larger than 2 MHz, and the injected current fails to entrain the oscillator

6.6 Spur-Approximation Region

As the detuning $\Delta\omega$ is further increased, the oscillator is very weakly disturbed from its free-running motion. Such behavior can be described as the almost free-running oscillator carrier at ω_o, accompanied by two dominant spurs located approximately at $\omega_o \pm \Delta\omega$. To obtain an expression for the frequency shift $\Delta\omega_{shift}$ of the weakly pulled carrier, we simplify (6.11) by setting $\Delta\omega \gg \alpha$. This gives

$$\phi(t) \approx \Delta\omega \cdot t - \sqrt{\Delta\omega^2 - \alpha^2} \cdot t \tag{6.14}$$

With the aid of (6.6), we obtain

$$\Delta\omega_{shift} = \omega_{osc} - \omega_o \approx \Delta\omega \cdot \left(1 - \sqrt{1 - \left(\frac{\alpha}{\Delta\omega}\right)^2} \right) \ll \Delta\omega \tag{6.15}$$

Equation (6.15) shows that the oscillator carrier is shifted by an amount $\Delta\omega_{shift}$ much smaller than the detuning $\Delta\omega$. In practice, we may assume that the oscillator enters the region of spur approximation for values of the detuning $|\Delta\omega| \geq 5\alpha$.

To obtain the level of the dominant spurs, we start from the phase dynamics equation (6.7), which we rewrite as shown below

$$\frac{d\phi(t)}{dt} = \frac{1}{A_1 C} \cdot \left[i_{inj_o} \cdot \cos[(\omega_o + \Delta\omega)t] \cdot \sin(\omega_{osc}t) \right] \tag{6.16}$$

where ω_{osc} is given by (6.15). Expanding (6.16) and integrating with respect to the time, we obtain

$$\phi(t) = \frac{i_{inj_o}}{2A_1\,C\,(\Delta\omega - \Delta\omega_{shift})} \cdot \cos[(\Delta\omega - \Delta\omega_{shift})t] \qquad (6.17)$$

The oscillator carrier becomes

$$v(t) = A_1 \cdot \cos[\omega_{osc}t + \phi(t)] \approx A_1 \cdot \cos(\omega_{osc}t) - A_1 \cdot \sin(\omega_{osc}t) \cdot \phi(t) \qquad (6.18)$$

Substituting (6.17) into (6.18) gives the two dominant spurs as

$$\frac{i_{inj_o}}{4\,C\,(\Delta\omega - \Delta\omega_{shift})} \cdot \sin[(\omega_o + \Delta\omega)t] \qquad (6.19)$$

and

$$\frac{i_{inj_o}}{4\,C\,(\Delta\omega - \Delta\omega_{shift})} \cdot \sin[(\omega_o - \Delta\omega + 2\Delta\omega_{shift})t] \qquad (6.20)$$

The two dominant spurs are not located symmetrically with respect to the unperturbed carrier frequency ω_o. Under the assumption of detuning much larger than the critical value $\Delta\omega \gg \alpha$, we have that $\Delta\omega \gg \Delta\omega_{shift}$ and $\omega_{osc} \approx \omega_o$. Equations (6.18), (6.19), and (6.20) simplify to

$$v(t) \approx A_1 \cdot \cos(\omega_o t) - \frac{i_{inj_o}}{4\,C\,\Delta\omega} \cdot \sin[(\omega_o \pm \Delta\omega)t] \qquad (6.21)$$

The oscillator motion is therefore approximated as the free-running carrier accompanied by two dominant spurs located approximately at $\omega_o \pm \Delta\omega$. The amplitude of the spurs is proportional to the amplitude of the injected current across the tank. It reduces with increasing tank capacitance and frequency offset from the carrier. Intuitively, we may consider that the injected current with amplitude i_{inj_o} at $\omega_o + \Delta\omega$ results in two spurs on either side of the carrier according to

$$\frac{1}{2}i_{inj_o}|Z(\omega_o \pm \Delta\omega)| \qquad (6.22)$$

where

$$|Z(\omega_o \pm \Delta\omega)| = \frac{1}{2C|\Delta\omega|} \qquad (6.23)$$

is the impedance of the self-sustained oscillator at offsets $\pm\Delta\omega$ from the carrier frequency as computed in (1.65). This is similar to the case of a current injected

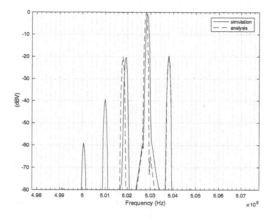

Fig. 6.12 Continuous curve: oscillator spectrum undergoing quasi-periodic motion (transient simulation). Dashed curve: oscillator spectrum undergoing quasi-periodic motion, where the oscillator phase is calculated by the approximative Eq. (6.21). $A_1 = 1$ V, $C = 1$ pF, $i_{inj_o} = 25$ uA, and the critical value of the detuning is $\alpha/(2\pi) = 2$ MHz. The detuning frequency is 10 MHz, and the oscillator spectrum can be approximated as a weakly pulled carrier with two dominant spurs

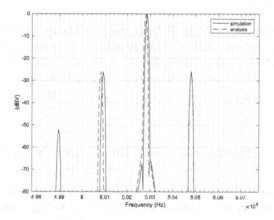

Fig. 6.13 Continuous curve: oscillator spectrum undergoing quasi-periodic motion (transient simulation). Dashed curve: oscillator spectrum undergoing quasi-periodic motion, where the oscillator phase is calculated by the approximative Eq. (6.21). $A_1 = 1$ V, $C = 1$ pF, $i_{inj_o} = 25$ uA, and the critical value of the detuning is $\alpha/(2\pi) = 2$ MHz. The detuning frequency is 20 MHz, and the oscillator spectrum can be approximated as a weakly pulled carrier with two dominant spurs

across a RLC tank discussed in Sect. 6.2, except that in the case of a self-sustained oscillator, the spurs appear on either side of the oscillator carrier.

In Figs. 6.12, 6.13, and 6.14, the continuous curves are obtained from transient simulation and show the oscillator spectrum depicted in Fig. 6.4 undergoing quasi-periodic motion. The dashed curves show the spectrum of the same oscillator, where the phase is obtained from the simplified Eq. (6.21). $A_1 = 1$ V, $C = 1$ pF, $i_{inj_o} = 25$ uA, and the critical value of the detuning is $\alpha/(2\pi) = 2$ MHz. The

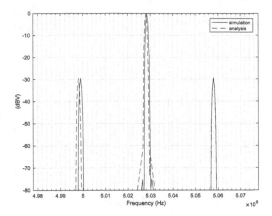

Fig. 6.14 Continuous curve: oscillator spectrum undergoing quasi-periodic motion (transient simulation). Dashed curve: oscillator spectrum undergoing quasi-periodic motion, where the oscillator phase is calculated by the approximative Eq. (6.21). $A_1 = 1$ V, $C = 1$ pF, $i_{inj_o} = 25$ uA, and the critical value of the detuning is $\alpha/(2\pi) = 2$ MHz. The detuning frequency is 30MHz, and the oscillator spectrum can be approximated as a weakly pulled carrier with two dominant spurs

detuning frequencies are 10 MHz, 20 MHz, and 30 MHz, respectively. The oscillator spectrum can be approximated as a weakly pulled carrier with two dominant spurs for such large detuning values. In all three figures, the simulated dominant spurs are not located symmetrically with respect to the carrier following the discussion above. Their level closely agrees with the value given in Eq. (6.21).

6.7 x2 LO Transmitter Architecture—Adler's Equation

Figure 6.15 shows a x2 LO transmitter architecture. The phase-locked loop (PLL) consists of a phase detector (PD), a loop filter (LF), a voltage-controlled oscillator (VCO), and a feedback divider (/N). The phase-locked loop generates at its output a local oscillator signal (LO), locked to the incoming reference signal (REF). The LO signal is divided by two before feeding the power amplifier (PA). Modulating the LO signal can be achieved by using a pair of mixers in series with the divide by two (DIV2) or by introducing the phase modulation in the PLL and the amplitude modulation in the PA. In either case, the PA second harmonic is at the same frequency as the carrier of the VCO. Depending on the PA output power, its second harmonic may be sufficiently strong to affect and pull the VCO. The PLL tries to correct the frequency shift, but it is generally not fast enough. This results in performance deterioration in modulation quality metrics such as error vector magnitude (EVM) and spectral regrowth.

Figure 6.16 shows the oscillator tank inductor L_1 aggressed by the second harmonic PA current i flowing in inductor L_2, which is part of the PA. The

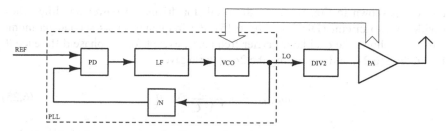

Fig. 6.15 Typical x2 LO transmitter architecture. The power amplifier's second harmonic is at the same frequency as the oscillator carrier, resulting in pulling the oscillator

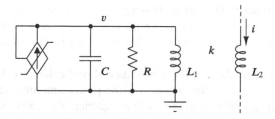

Fig. 6.16 The power amplifier's second harmonic current i flowing in the PA inductor L_2 aggresses the oscillator tank inductor L_1. $k = M/\sqrt{L_1 L_2}$ is the magnetic coupling factor between L_1 and L_2. M is the mutual inductance

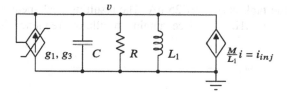

Fig. 6.17 Equivalent model of the magnetic coupling between the power amplifier inductor L_2 and the oscillator inductor L_1. i is the power amplifier's second harmonic flowing in the inductor L_2. M is the mutual coupling between inductors L_1 and L_2. Terms g_1 and g_3 model the nonlinear current source according to (2.16)

arrangement in Fig. 6.16 can be modeled as is shown in Fig. 6.17, where M is the mutual inductance between L_1 and L_2.

Similarly to (6.6) and (6.7), in order to estimate the frequency shift $\omega_{shift} = \omega_{osc} - \omega_o$ of the oscillator carrier $v = A_1 \cos(\omega_{osc} t)$ from its free-running frequency ω_o, we use the phase dynamics equation

$$\frac{d\phi(t)}{dt} = \frac{1}{A_1 C} \cdot \left[i_{inj_o} \cdot \cos(\omega_{osc} t + \theta) \right] \cdot \sin(\omega_{osc} t) \tag{6.24}$$

In (6.24), i_{inj_o} is the amplitude of the injected current i_{inj} across the oscillator tank depicted in the equivalent model in Fig. 6.17. θ is the phase difference between the power amplifier's second harmonic current i flowing in L_2 and the oscillator carrier, as shown in Fig. 6.16. As the oscillator drives the power amplifier via a divide-by-two circuit (DIV2 in Fig. 6.15), the frequency of its second harmonic current i follows the oscillator carrier frequency, except for the phase difference θ. Expanding (6.24) and discarding the $2\omega_{osc}$ term give

$$\omega_{shift} = -\frac{i_{inj_o}}{2\,A_1\,C} \cdot \sin\theta \qquad (6.25)$$

The frequency shift is proportional to the amplitude of the injected current i_{inj_o} across the tank and inversely proportional to the oscillator carrier amplitude A_1 and the tank capacitance C. Furthermore, it depends on the phase θ of the power amplifier's second harmonic current flowing in L_2 relative to the oscillator carrier. The frequency shift is minimized if the injected current is in-phase or anti-phase with the oscillator carrier. Equation (6.25) is known in the literature as Adler's equation [23].

Figure 6.18 depicts the oscillator's simulated and calculated frequency shifts shown in Fig. 6.16 against the phase θ of the power amplifier second harmonic current flowing in L_2 relative to the oscillator carrier. The continuous curve is the simulated frequency shift, while the dashed curve is the estimated frequency shift according to Eq. (6.25). We have set $g_1 = 5$ mS, $g_3 = 4$ mA/V^3, $R = 500$ Ohm for the oscillator. These values give oscillator amplitude equal to $A_1 = 1$ V. The tank components are $C = 1$ pF, $L = 1$ nH, while the amplitude of the injected current across the tank is $i_{inj_o} = 25$ uA. The resulting peak–peak frequency shift is approximately 4 MHz. The free-running oscillator frequency is estimated by

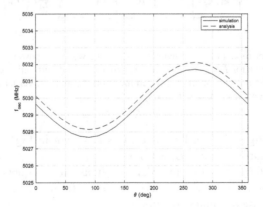

Fig. 6.18 Continuous curve: Simulation of the oscillator depicted in Fig. 6.16, showing the oscillator frequency shift against the phase θ of the power amplifier's second harmonic current flowing in L_2, relative to the oscillator carrier. Dashed curve: Estimation of the oscillator frequency shift using Eqs. (2.57), (2.58), and (6.25). In this example, $g_1 = 5$ mS, $g_3 = 4$ mA/V^3, $R = 500$ Ohm. These values give oscillator amplitude equal to $A_1 = 1$ V. The tank components are $C = 1$ pF, $L = 1$ nH, while the amplitude of the injected current across the tank is $i_{inj_o} = 25$ uA. The resulting peak–peak frequency shift is approximately 4 MHz, in close agreement with the simulated value

$\omega_o = 1/\sqrt{LC}$ with a correction factor $(1 - \epsilon^2/16)$ as derived in (2.57) and (2.58). Very good agreement between the simulated results and analysis is observed.

Pulling of the VCO can be detrimental in applications where the modulation format entails amplitude modulation. This is because the frequency shift is dependent on the fast variations of the modulated carrier envelope. As mentioned above, the PLL tries to correct the frequency shift, but it is generally not fast enough. This results in performance deterioration in modulation quality metrics such as EVM and spectral regrowth. Equation (6.25) reveals that to mitigate the problem, we need to reduce the equivalent current injected across the tank. Therefore, we need to reduce the power amplifier's second harmonic current and minimize the magnetic coupling between the power amplifier and the oscillator tank. Techniques for reducing magnetic coupling are discussed in this chapter. Additionally, we may resort to methods that control the value of the phase shift θ between the oscillator carrier and the power amplifier's second harmonic current so that to operate under the condition where $\sin \theta$ is approximately zero in (6.25).

As an example, we apply Eq. (6.25) to estimate the frequency pulling due to the magnetic coupling between the power amplifier and the oscillator depicted in Figs. 6.15, 6.16 and 6.17. Assuming that the power amplifier's second harmonic current amplitude is $i_o = 10$ mA, the oscillator tank components are $L_1 = 1$ nH and $C = 1$ pF (5 GHz center frequency), and the oscillator amplitude is $A_1 = 1$ V, the tiny amount of mutual coupling $M = 2.5$ pH between the power amplifier inductor L_2 and the oscillator inductor L_1 results in approximately 4 MHz peak–peak pulling. Minute amounts of magnetic coupling result in sizeable frequency shifts that can harm digital modulation formats. In such cases, digital cancelation methods are indispensable.

6.8 Magnetic Coupling Between Loops

In this second part of the chapter, we concentrate on magnetic coupling and discuss ways to reduce it. Magnetic coupling from the power amplifier to the oscillator in Fig. 6.15 may happen from the power amplifier core or supply/ground interconnections. In this section, we concentrate on the former phenomenon and discuss the latter in the following section.

Let us assume a circular planar loop with radius a carrying current I as is shown in Fig. 6.19. We want to estimate the magnetic flux density B at point P on the loop plane at a distance x from the loop center, where $x \gg a$.

We apply the Biot–Savart law along the perimeter of the circular loop [24]

$$B(x) = \frac{\mu_o I}{4\pi} \oint \frac{dl \times \hat{r}}{r^2} = \frac{\mu_o I}{4\pi} \int_0^{2\pi} \frac{a\, d\theta \sin \psi}{r^2} \qquad (6.26)$$

Applying twice the law of cosines in Fig. 6.19 gives the pair of equations

Fig. 6.19 In estimating the
magnetic field at large
distances from a
current-carrying loop, it
suffices to treat the loop as a
magnetic dipole

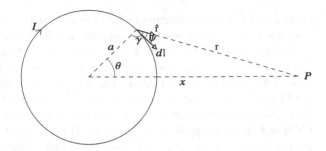

$$x^2 = a^2 + r^2 - 2ar \cos \gamma \qquad (6.27)$$

$$r^2 = a^2 + x^2 - 2ax \cos \theta \qquad (6.28)$$

substituting $\cos \gamma = -\sin \psi$ since $\gamma = \pi/2 + \psi$, and using that $x \gg a$, the pair of
equations reduces to

$$x^2 = a^2 + r^2 + 2ar \sin \psi \qquad (6.29)$$

$$r^2 \approx x^2 - 2ax \cos \theta \qquad (6.30)$$

These two equations allow us to express the ratio $\sin \psi / r^2$ in Eq. (6.26) in terms of
$a, x,$ and θ as follows:

$$B(x) \approx \frac{\mu_o I a}{4\pi x^2} \int_0^{2\pi} \frac{\cos \theta - \frac{a}{x}}{\left(1 - \frac{2a}{x} \cos \theta\right)^{3/2}} d\theta \qquad (6.31)$$

Since $x \gg a$, (6.31) can be further approximated as

$$B(x) \approx \frac{\mu_o I a}{4\pi x^2} \int_0^{2\pi} \left(\cos \theta - \frac{a}{x} \right)\left(1 + \frac{3a}{x} \cos \theta \right) d\theta \qquad (6.32)$$

The integration in (6.32) is straightforward giving

$$B(x) \approx \frac{\mu_o I A}{4\pi x^3} \qquad (6.33)$$

The product of the loop current I times the loop area $A = \pi a^2$ in the numerator
of (6.33) is known as the magnetic dipole moment [24]. Therefore, the magnetic flux
density $B(x)$ is proportional to the magnetic dipole moment and decays as $1/x^3$.
This approximation is the magnetic dipole approximation. It can be utilized in
all cases where we are interested in evaluating the magnetic flux density at large
distances from a loop. The direction of the magnetic flux density vector at point P
in Fig. 6.19 points outside the page, as the right-hand rule requires.

Fig. 6.20 Magnetic flux density B at point P, located on the plane of a circular loop, at a distance x from the loop center as depicted in Fig. 6.19. The loop radius is $a = 100$ μm, while the circulating current is $I = 1$ mA. Continuous curve: Numerical integration of Eq. (6.26). Dashed curve: Prediction of approximate Eq. (6.33)

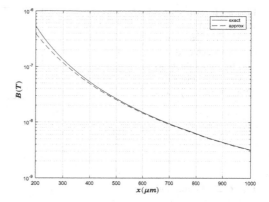

It is instructive to verify the limitations imposed by the approximation $x \gg a$ in deriving (6.33). In Fig. 6.20, we compare Eq. (6.33) with the result obtained by numerical integration of (6.26). The loop radius is $a = 100$ μm, while the circulating current is $I = 1$ mA. Equation (6.33) is a very good approximation, even for points relatively close to the loop.

We can use Eq. (6.33) to obtain an estimate of the mutual inductance between two circular loops that are spaced apart by distance x as follows. The magnetic flux threading loop 2 due to the current I circulating in loop 1 is written as

$$\Phi_{21} \approx B\,A_2 \approx \frac{\mu_o\,I\,A_1\,A_2}{4\pi x^3} \tag{6.34}$$

x is the center-to-center separation of the loops. We have assumed that the magnetic flux density B threading loop 2 is approximately constant and equal to its value at the center of loop 2 as given by (6.33). Since $\Phi_{21} = MI$ [24], it follows that

$$M_{21} \approx \frac{\mu_o\,A_1\,A_2}{4\pi x^3} \tag{6.35}$$

The mutual inductance M_{21} in (6.35) has been estimated for two single-turn circular loops. For loops with multiple turns, we need to multiply the expression in (6.35) by the number of turns of each coil.

To validate the usefulness of (6.35), we apply it to evaluate the mutual inductance between two square-shaped loops. A_1, A_2 are the areas of the loops, while x is their center-to-center separation as is shown in Fig. 6.21. The predicted values of the mutual inductance are compared to electromagnetic simulation in Fig. 6.22. For center-to-center separations larger than 2–3 times the loop edge lengths, Eq. (6.35) approximates the mutual coupling well.

Equation (6.35) shows the strong dependence of the magnetic coupling between loops on their separation. Mutual coupling is inversely proportional to the third power of the loop center-to-center distance. In most practical applications, the loop areas A_1 and A_2 are determined by performance metrics other than magnetic

Fig. 6.21 Two square loops with areas A_1 and A_2, respectively, separated by center-to-center distance x. The mutual inductance between the loops is approximated by Eq. (6.35)

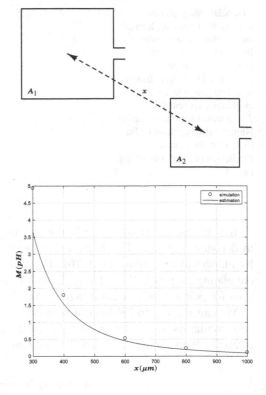

Fig. 6.22 Mutual inductance between two planar square loops with side length $l = 177$ μm as a function of their center-to-center separation. Circles: electromagnetic simulation. Continuous curve: Eq. (6.35)

coupling. Depending on the application, even with large separations, the magnetic coupling may result in interactions that can be detrimental. Later, we discuss methods to reduce the magnetic coupling between loops.

6.9   Magnetic Coupling Between Line and Loop

As discussed in the previous section, magnetic coupling from the power amplifier to the oscillator in Fig. 6.15 may also happen from the power amplifier supply/ground interconnections. In an x2 LO transmitter architecture, such lines may carry significant energy at the power amplifier's second harmonic, thus aggressing the oscillator. This magnetic coupling can be approximated by the coupling between an infinite line and a loop, as shown in Fig. 6.23.

To estimate the magnetic coupling, we start from Ampere's law, giving the magnetic flux density at a distance x from the infinite line carrying current I as [24]

$$B = \frac{\mu_o I}{2\pi x} \tag{6.36}$$

Fig. 6.23 Square-shaped
loop with dimensions a, b,
separated from an infinite line
by distance c. The mutual
inductance between the
infinite line and the loop is
given by Eq. (6.38)

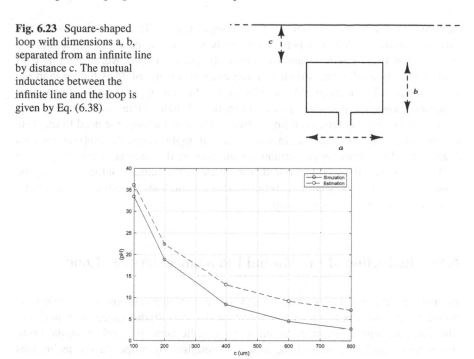

Fig. 6.24 Continuous curve: electromagnetic simulation of mutual inductance between a squared-shaped loop with dimensions $a = b = 177$ μm and a line of length 885 μm as a function of their separation c. The arrangement is shown in Fig. 6.23. Dashed curve: estimation of mutual inductance between the same squared-shaped loop and an infinite line as a function of their separation c, according to Eq. (6.38)

The magnetic flux Φ threading the squared-shaped loop due to the current I flowing in the infinite line is written as

$$\Phi = \frac{\mu_o I}{2\pi} \int_c^{c+b} \frac{a\,dx}{x} = \frac{\mu_o I\,a}{2\pi} \cdot ln\left(\frac{b+c}{c}\right) \tag{6.37}$$

Since $\Phi = M\,I$, it follows that the mutual inductance between the infinite line and the squared-shaped loop in Fig. 6.23 is given by

$$M = \frac{\mu_o\,a}{2\pi} \cdot ln\left(1 + \frac{b}{c}\right) \tag{6.38}$$

The mutual inductance M in (6.38) has been estimated for a single-turn squared-shaped loop. We must multiply the expression in (6.38) for a multiple-turn loop by the turn number.

In Fig. 6.24, the continuous curve shows the simulated mutual inductance between a squared-shaped loop with dimensions $a = b = 177$ μm and a line of length 885 µm as a function of their separation c. The dashed curve is the estimated mutual inductance between the same squared-shaped loop and an infinite line as a function of their separation c, according to Eq. (6.38). Because a line with finite length is used in the simulation, the estimated value of M is higher than the simulated one. Equation (6.38), therefore, quantifies the upper bound of the mutual coupling between a loop and a line as a function of their separation c.

To reduce the magnetic coupling between a line and a loop, we need to increase the distance c. This is because, in most practical applications, the loop dimensions a and b are determined by performance metrics other than magnetic coupling. Even with separations of several hundred micrometers, the mutual inductance may be unacceptably large. In a later section, we discuss methods to reduce the magnetic coupling between a line and a loop.

6.10 Reduction of the Mutual Inductance Between Loops

Improving inductor magnetic immunity is vital in applications where magnetic coupling results in performance degradation. Even if digital cancelation and pre-distortion techniques are used to meet the specifications of modern digital communication standards, minimizing magnetic coupling by proper layout techniques is usually the first step that needs to be taken. In this section, we concentrate on the figure-eight (fig8-shaped) inductor structure to increase the magnetic immunity of the victim oscillator inductor. We also discuss routing supply/ground and clock aggressor lines in magnetically immune pairs to minimize their magnetic radiation.

Figure 6.25 shows a squared-shaped loop with area A_1 and a fig8-shaped loop. Each lobe of the fig8 structure has an area $A_2/2$. The center-to-center separations between the squared-shaped loop and each of the lobes of the fig8-shaped loop are x_1 and x_2, respectively. Using Eq. (6.35) and taking into account the opposing polarities of the two lobes that constitute the fig8 structure, we obtain the mutual inductance

$$M \approx \frac{\mu_o A_1 A_2}{8\pi} \left(\frac{1}{x_1^3} - \frac{1}{x_2^3} \right) \tag{6.39}$$

Equation (6.39) shows that as x_1 and x_2 increase, the mutual inductance between the squared-shaped loop and the fig8-shaped loop diminishes. This is due to the opposing polarities of the fig8-shaped loop lobes.

The continuous curve in Fig. 6.26 shows the simulated mutual inductance between a squared and a fig8-shaped loop as a function of their separation x. Both loops have dimensions $a = b = 177$ µm. The arrangement is shown in the lower part of Fig. 6.27. The dashed curve shows the simulated mutual inductance between two squared-shaped loops as a function of their separation x. Both loops have dimensions $a = b = 177$ µm. The arrangement is shown in the upper part of Fig. 6.27. The magnetic coupling to the fig8-shaped loop is reduced by a factor of

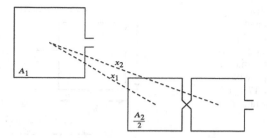

Fig. 6.25 A square-shaped loop with area A_1 and a fig8-shaped loop with total area A_2. The separations between the squared-shaped loop and the two lobes of the fig8-shaped loop are x_1 and x_2, respectively. The mutual inductance between the two structures is approximately given by Eq. (6.39)

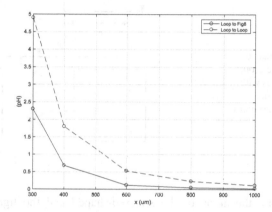

Fig. 6.26 Continuous curve: Electromagnetic simulation of mutual inductance between a squared and a fig8-shaped loop as a function of their separation x. Both loops have the same dimensions $a = b = 177$ μm. The arrangement is shown in the lower part of Fig. 6.27. Dashed curve: Electromagnetic simulation of mutual inductance between two squared-shaped loops as a function of their separation x. Both loops have the same dimensions $a = b = 177$ μm. The arrangement is shown in the upper part of Fig. 6.27

two for short separations and up to a factor of five for large separations. Therefore, as the loop separation increases, the mutual inductance between a squared-shaped loop and a fig-8 shaped loop reduces faster than in the case of two squared-shaped loops. Let us consider Fig. 6.28 to see this.

The mutual inductance between the squared-shaped loop and the fig8-shaped loop can be approximated by applying Eq. (6.35) to each lobe of the fig8-shaped inductor

$$M \approx \frac{\mu_o A_1 \frac{A_2}{2}}{4\pi}\left(\frac{1}{(x - \frac{b}{4})^3} - \frac{1}{(x + \frac{b}{4})^3}\right) \qquad (6.40)$$

Fig. 6.27 Top part: two
square-shaped loops with
dimensions $a = b = 177\,\mu$m.
Center-to-center separation is
x. Bottom part: square-shaped
and fig8-shaped loops, with
dimensions $a = b = 177\,\mu$m.
Center-to-center separation
is x

Fig. 6.28 Square-shaped
loop and fig8-shaped loop.
The mutual decreases with
the fourth power of the
center-to-center separation x

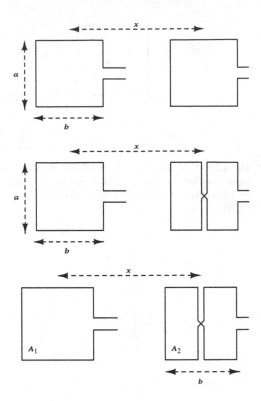

where A_1, A_2 are the areas of the squared-shaped and the fig8-shaped loops, respectively. For large separations $(x \gg b)$, Eq. (6.40) can be simplified as

$$M \approx \frac{\mu_o A_1 A_2}{4\pi x^3} \cdot \left(\frac{3b}{4x}\right) \tag{6.41}$$

The first term in (6.41) is the same as Eq. (6.35) and gives the mutual inductance between two squared-shaped loops with areas A_1 and A_2. The coupling reduces with the third power of the center-to-center separation. The additional term in parenthesis shows that if one of the loops is replaced with a fig8-shaped structure, the mutual inductance becomes inversely proportional to the fourth power of the loop separation.

To further reduce the mutual coupling between loops, we may consider the orientation of fig8-shaped loops. Figure 6.29 shows two different orientations for the fig8-shaped loop. The dimensions of the loops are $a = b = 177\,\mu$m, while center-to-center separation is $x = 600$ um. In the lower part of the figure, the fig8-shaped loop lobes are symmetrically placed with respect to the squared-shaped aggressor loop. Considering Eq. (6.39), such an orientation results in approximately the same values for the separations between the aggressor center and each lobe. As a result, the mutual inductance is expected to be significantly reduced. EM simulation of the

Fig. 6.29 Square-shaped
loop and fig8-shaped loop
with dimensions
$a = b = 177\,\mu$m.
Center-to-center separation is
$x = 600$ um. The orientation
of the fig8-shaped loop at the
bottom part of the figure is
symmetrical with respect to
the square-shaped loop and
helps minimize the mutual
coupling

Fig. 6.30 Infinite line and
fig8-shaped loop. The fig8
structure has dimensions a
and b. Its two lobes are
separated from the infinite
line by distances c_1 and c_2,
respectively

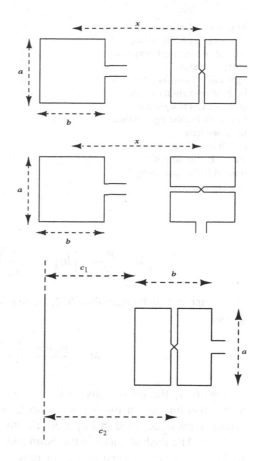

orientation on the top part of Fig. 6.29 gives $M = 118$ fH, while that on the bottom
offers a fourfold reduction, giving $M = 30$ fH.

6.11   Reduction of the Mutual Inductance Between a Line and a Loop

Power amplifiers' supply and ground metallization may carry significant energy at
the second harmonic. Such lines may aggress the oscillator in x2 LO architectures
depicted in Fig. 6.15. This section discusses using fig8-shaped structures to reduce
the mutual inductance between a line and a loop.

Figure 6.30 shows an infinite line and a fig8-shaped loop. The fig8 structure
has dimensions a and b, and each of its lobes is separated from the infinite line
by distances c_1 and c_2, respectively. To estimate the mutual inductance, we apply
Eq. (6.38) on each lobe separately giving

Fig. 6.31 Continuous curve: electromagnetic simulation of mutual inductance between a 900 μm line and a fig8-shaped loop as a function of their separation c_1. The arrangement is shown in Fig. 6.30. For the fig8-shaped loop, we have $a = b = 177$ μm. Dashed curve: Estimation of the mutual inductance using Eq. (6.43)

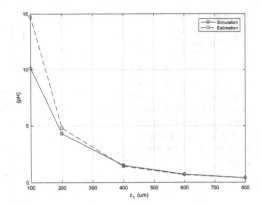

$$M = \frac{\mu_o\,a}{2\pi} \cdot \left[ln\left(1 + \frac{b}{2c_1}\right) - ln\left(1 + \frac{b}{2c_2}\right) \right] \qquad (6.42)$$

For c_1 and c_2 much larger than $b/2$, we may use the approximation $ln(1 + x) \approx x$ giving

$$M \approx \frac{\mu_o\,a\,b}{2\pi} \cdot \frac{1}{2} \cdot \left(\frac{1}{c_1} - \frac{1}{c_2}\right) \qquad (6.43)$$

In Fig. 6.31, the continuous curve is the simulated mutual inductance between a 900 μm line and a fig8-shaped loop as a function of their separation c_1. The arrangement is depicted in Fig. 6.30. For the fig8-shaped loop, we have $a = b = 177$ μm. The dashed curve is the estimated value of the mutual inductance based on Eq. (6.43). As the separation c_1 increases, the agreement between Eq. (6.43) and simulation improves.

To demonstrate that using a fig8-shaped structure results in reduced mutual inductance compared to using a simple loop, we start from (6.43) and set $c_2 = c_1 + b/2$. This gives

$$M \approx \frac{\mu_o\,a\,b}{2\pi} \cdot \frac{1}{2} \cdot \left(\frac{1}{c_1} - \frac{1}{c_1 + b/2}\right) \qquad (6.44)$$

For $c_1 \gg b/2$, Eq. (6.44) reduces to

$$M \approx \frac{\mu_o\,a\,b}{2\pi c_1} \cdot \left(\frac{b}{4c_1}\right) \qquad (6.45)$$

The first term in (6.45) corresponds to the mutual inductance between an infinite line and a loop with dimensions a, b, separated by distance c_1. The mutual inductance is inversely proportional to the separation. The term in parenthesis captures the additional reduction in mutual inductance due to the geometry of the

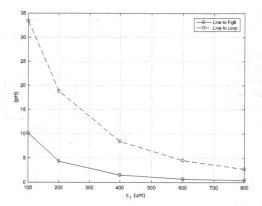

Fig. 6.32 Continuous curve: Electromagnetic simulation of mutual inductance between a 900 μm line and a fig8-shaped loop as a function of their separation c_1. Dashed curve: Electromagnetic simulation of mutual inductance between a 900 μm line and a square-shaped loop as a function of their separation c_1. Both loops have dimensions $a = b = 177$ μm. The length of the line is 900 μm

fig8-shaped loop. The mutual inductance diminishes faster and becomes inversely proportional to the square of the separation. This is demonstrated in Fig. 6.32, where the continuous curve corresponds to the fig8-shaped loop case, while the dashed curve corresponds to the square loop case. Both loops have dimensions $a = b = 177$ μm, while the line length is 900 μm.

To further reduce the mutual coupling, we may consider the orientation of fig8-shaped loops. In the arrangement depicted in Fig. 6.30, the simulated mutual inductance between the 900 μm long line and the fig8-shaped loop with dimensions $a = b = 177$ μm is $M = 740$ fH for a separation $c_1 = 600$ μm. Rotating the fig8-shaped loop by 90 degrees reduces the mutual coupling threefold to $M = 235$ fH. This is because the two opposing lobes of the fig8-shaped structure are placed symmetrically with respect to the line, increasing their cancelation.

Low-frequency clock lines often exhibit sharp transitions resulting in significant energy at high frequencies, which couple magnetically to inductors in sensitive transceiver parts, such as the oscillator inductor, the inductors in the front end of the receiver, etc. In the last part of this section, we discuss the routing of aggressing lines in magnetically immune pairs to minimize their magnetic radiation. To achieve this, we route the lines very close to each other and arrange them so that the aggressing currents flow in opposite directions.

Let us consider the arrangement shown in Fig. 6.33. A square-shaped loop is separated by an infinite pair of lines with opposing currents by distance c. The lines are spaced apart by Δc. The loop has dimensions a and b. To calculate the mutual inductance, we apply Eq. (6.38) for each line giving

$$M = \frac{\mu_o a}{2\pi} \cdot \left[ln\left(1 + \frac{b}{c - \Delta c}\right) - ln\left(1 + \frac{b}{c}\right) \right] \qquad (6.46)$$

Fig. 6.33 Infinite pair of lines and squared-shaped loop. The loop has dimensions a and b and is separated from the infinite lines by distance c. The pair of lines is spread apart by Δc. The currents flowing in the lines are in opposing directions

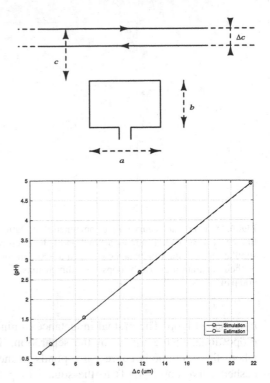

Fig. 6.34 Continuous curve: electromagnetic simulation of mutual inductance between two 900 μm long lines and a squared-shaped loop as a function of the pair separation Δc. Dashed curve: Estimation of the mutual inductance according to (6.48). The loop has dimensions $a = b = 177$ μm. The distance from the pair of lines to the loop is 100 μm

To proceed, we use that

$$1 + \frac{b}{c - \Delta c} = 1 + \frac{b}{c(1 - \frac{\Delta c}{c})} \approx 1 + \frac{b}{c} + \frac{b}{c^2}\Delta c \qquad (6.47)$$

Substituting (6.47) into (6.46) gives

$$M = \frac{\mu_o a}{2\pi} \cdot ln\left(1 + \frac{b}{c} \cdot \frac{\Delta c}{(c + b)}\right) \approx \frac{\mu_o a b}{2\pi c} \cdot \frac{\Delta c}{(c + b)} \qquad (6.48)$$

The approximation in (6.48) is valid for adequately small Δc. M diminishes as Δc goes to zero. In Fig. 6.34, the continuous curve is the simulated mutual inductance between two 900 μm long lines and a squared-shaped loop as a function of the pair separation Δc. The dashed curve is the estimated mutual inductance according to (6.48). The loop has dimensions $a = b = 177$ μm. The distance from the pair of lines to the loop is 100 μm. A very good agreement between simulation and estimation is observed.

In the remaining chapter of this book, we concentrate on the design of integrated inductors. We approach the subject from the ground up to explain the basic principles behind inductor design. We build a simple model that captures the essential effects that determine inductor performance, such as inductance value and loss mechanisms. As we have done in the rest of this book, we emphasize approximate calculations that highlight the physical phenomena and provide design insight as a complement to simulations.

Chapter 7
Design of Integrated Inductors

7.1 Introduction

The realization of fully integrated LC oscillators depends on the availability of good-quality on-chip passive components. In particular, integrated inductors are indispensable for realizing LC oscillators operating in the GHz range. Good-quality inductors are challenging to realize monolithically due to thin metals and capacitive parasitics impairing their performance. The main goal of inductor design is maximizing inductance value and quality factor while minimizing area and possibly coupling.

In this chapter, we discuss issues pertaining to the analysis and design of planar spiral inductors for fully integrated VCOs. We build a simple model that captures the essential effects that determine inductor performance, such as inductance value and loss mechanisms. We emphasize approximate calculation methods that provide design insight as a complement to electromagnetic simulations.

7.2 Mutual Inductance of Filamentary Loops

Let us consider two filamentary plane loops, A, and B, of arbitrary shapes in free space as is shown in Fig. 7.1. The voltage V_{ind} induced in loop B due to current I circulating in loop A is given by Faraday's law [25]

$$V_{ind} = -\frac{d\Phi_B}{dt} \tag{7.1}$$

Equation (7.1) states that the induced voltage V_{ind} in loop B is proportional to the rate of change of the magnetic flux Φ_B threading it. The minus sign indicates that the polarity of V_{ind} is such that it opposes the change of the magnetic flux as required

© The Author(s), under exclusive license to Springer Nature Switzerland AG 2023
K. Manetakis, *Topics in LC Oscillators*,
https://doi.org/10.1007/978-3-031-31086-7_7

Fig. 7.1 The current in loop
A sets up a magnetic field in
loop B. The rate of change of
the magnetic flux through
loop B induces a voltage V_{ind}
in that loop

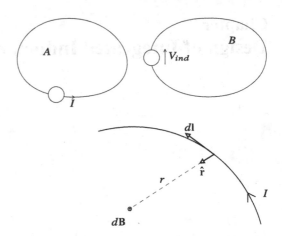

Fig. 7.2 The magnetic flux
density $d\mathbf{B}$ due to an
infinitesimal current element
$I d\mathbf{l}$ is given by the
Biot–Savart law. $d\mathbf{B}$ is normal
to the plane defined by $d\mathbf{l}$ and
$\hat{\mathbf{r}}$ and its direction is given by
the right-hand rule

by Lenz's law. The magnetic flux threading loop B is given by the surface integral
of the magnetic flux density $\mathbf{B}$ over any surface bounded by loop B.

$$\Phi_B = \iint \mathbf{B} \cdot d\mathbf{S} \qquad (7.2)$$

In (7.2), $d\mathbf{S} = dS \cdot \hat{\mathbf{n}}$, where $\hat{\mathbf{n}}$ is the normal unit vector at each point on the surface
over which the integral is evaluated.

To evaluate the magnetic flux density $\mathbf{B}$ at some point in space due to a
filamentary loop of arbitrary shape carrying constant current I, we subdivide the
loop into infinitesimal current elements and sum the individual contributions. The
contribution from each infinitesimal current element is given by the Biot–Savart law
[25]

$$d\mathbf{B} = \frac{\mu_o I}{4\pi} \frac{d\mathbf{l} \times \hat{\mathbf{r}}}{r^2} \qquad (7.3)$$

where the unit vector $\hat{\mathbf{r}}$ points from the current element $I d\mathbf{l}$ to the point where we
calculate the field contribution, and r is the corresponding distance as is shown in
Fig. 7.2. The constant of proportionality μ_o is called the permeability of free space
and is equal to $4\pi \, 10^{-7} \, H/m$.

The magnetic flux density at every point in space due to the current I in loop A
can thus be evaluated by the line integral around loop A

$$\mathbf{B} = \frac{\mu_o I}{4\pi} \oint \frac{d\mathbf{l}_A \times \hat{\mathbf{r}}}{r^2} \qquad (7.4)$$

Equation (7.4) shows that the magnetic flux density is proportional to the current
flowing in loop A. Therefore, so too is the magnetic flux Φ_B through loop B

$$\Phi_B = MI \tag{7.5}$$

The constant of proportionality is called the mutual inductance M and depends only on the geometry of the two loops [25].

To estimate the mutual inductance M between the two filamentary loops, therefore, we need first to calculate the magnetic flux density $\mathbf{B}$ at every point on a surface that is bounded by loop B by carrying out the line integral (7.4) and subsequently to calculate the magnetic flux Φ_B threading loop B by carrying out the surface integral (7.2). We thus have

$$M = \frac{\mu_o}{4\pi} \iint \left(\oint \frac{d\mathbf{l}_A \times \hat{\mathbf{r}}}{r^2} \right) d\hat{\mathbf{S}}_B \tag{7.6}$$

The calculation in (7.6) is computationally intensive. A more efficient way is to circumvent the intermediate calculation of the magnetic flux density $\mathbf{B}$ and instead calculate the magnetic flux Φ_B directly from the current I flowing in loop A. To do this, we use the notion of the vector magnetic potential $\mathbf{A}$ [26].

Maxwell's second equation states that there are no isolated magnetic monopoles. Therefore, the surface integral of the magnetic flux density over a closed surface is zero

$$\oiint \mathbf{B} \cdot d\mathbf{S} = 0 \tag{7.7}$$

Using the divergence theorem, we can thus write for the magnetic flux density vector [27]

$$\nabla \cdot \mathbf{B} = 0 \tag{7.8}$$

Equations (7.7) and (7.8) are known as the integral and differential forms of Maxwell's second equation, respectively. Since the divergence of the curl of a vector is zero, (7.8) invites the introduction of a vector field $\mathbf{A}$ given by [26]

$$\mathbf{B} = \nabla \times \mathbf{A} \tag{7.9}$$

The advantage of writing the magnetic flux density like this is that any vector field $\mathbf{A}$ satisfies (7.8). We may therefore choose $\mathbf{A}$ to simplify the problem at hand.

Ampère's law states that the line integral of the magnetic flux density $\mathbf{B}$ around a closed path in free space is proportional to the current through any surface that is bounded by the path

$$\oint \mathbf{B} \cdot d\mathbf{l} = \mu_o I \tag{7.10}$$

Using Stoke's theorem, we can thus write [27]

$$\nabla \times \mathbf{B} = \mu_o \cdot \mathbf{J} \tag{7.11}$$

where $\mathbf{J}$ is the current density vector. Substituting (7.9) into (7.11), we obtain

$$\nabla \times (\nabla \times \mathbf{A}) = \mu_o \cdot \mathbf{J} \tag{7.12}$$

which becomes

$$\nabla \cdot (\nabla \cdot \mathbf{A}) - \nabla^2 \cdot \mathbf{A} = \mu_o \cdot \mathbf{J} \tag{7.13}$$

To simplify (7.13), we require that $\mathbf{A}$ has zero divergence $\nabla \cdot \mathbf{A} = 0$. Equation (7.13) then becomes

$$\nabla^2 \mathbf{A} = -\mu_o \cdot \mathbf{J} \tag{7.14}$$

which is Poisson's equation. With the assumption that $\mathbf{A}$ goes to zero at infinity, the solution to (7.14) becomes [26]

$$\mathbf{A} = \frac{\mu_o I}{4\pi} \oint \frac{d\mathbf{l}}{r} \tag{7.15}$$

where the integral is taken around the loop that creates the field. The vector field $\mathbf{A}$ is called the vector magnetic potential, and we may regard (7.15) as the equivalent of (7.4). Figure 7.3 shows that the vector magnetic potential $d\mathbf{A}$ due to an infinitesimal current element $I \cdot d\mathbf{l}$ has the same direction as vector $d\mathbf{l}$. The vector magnetic potential at any point in space is the sum of the contributions from each infinitesimal current element $I \cdot d\mathbf{l}$.

Stoke's theorem can now be invoked to calculate the magnetic flux directly from the current

$$\Phi_B = \iint \mathbf{B} \cdot d\mathbf{S_B} = \oint \mathbf{A} \cdot d\mathbf{l}_B \tag{7.16}$$

Fig. 7.3 The vector magnetic potential $d\mathbf{A}$ due to an infinitesimal current element $I \cdot d\mathbf{l}$ has the same direction as vector $d\mathbf{l}$

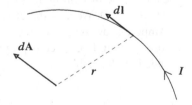

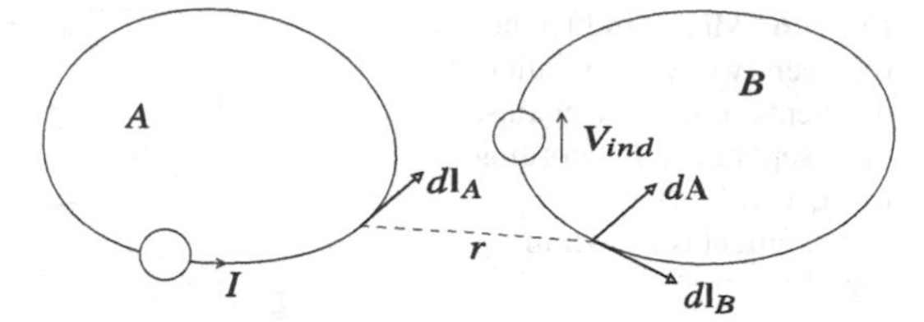

Fig. 7.4 Mutual inductance estimation has been reduced to the evaluation of two line integrals given in Eq. (7.17)

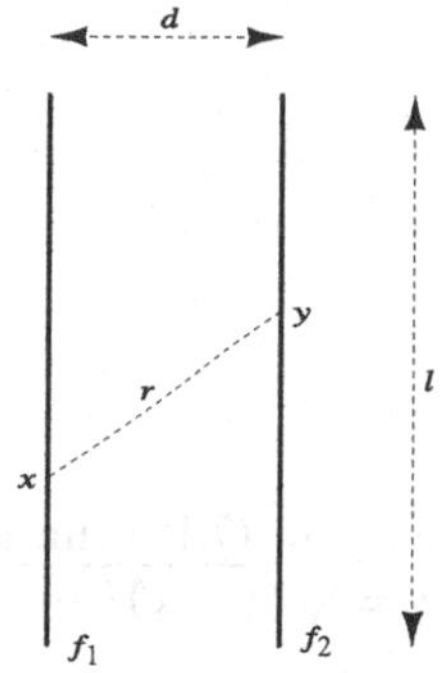

Fig. 7.5 Two parallel filaments f_1 and f_2, respectively, each of length l, are separated by distance d. Application of Neumann's form results in Eq. (7.21)

This allows us to replace the surface integral over a surface bounded by loop B with a line integral along that loop. The estimation of the mutual inductance reduces to the evaluation of two line integrals as is shown in Fig. 7.4

$$M = \frac{\mu_o}{4\pi} \oint d\mathbf{l}_B \oint \frac{d\mathbf{l}_A}{r} \qquad (7.17)$$

In (7.17), the inner line integral runs on loop A and estimates the vector magnetic potential **A** at every point on loop B, while the outer line integral runs on loop B and estimates the magnetic flux threading that loop.

Equation (7.17) is known as Neumann's form [26]. It reveals that mutual inductance is a purely geometrical quantity. Indeed, the integral in (7.17) depends only on μ_o and the geometry of the loops. Furthermore, it remains unchanged if we switch the roles of loops A and B. The induced voltage in loop B due to a current changing in loop A is the same as in loop A due to the same current changing in loop B. The adjective "mutual" emphasizes this property.

7.3 Mutual Inductance of Parallel Filaments

Let us assume two parallel filaments f_1 and f_2, of equal lengths l, separated by distance d as is shown in Fig. 7.5. Application of Neumann's form results in the following double integral:

$$M = \frac{\mu_o}{4\pi} \int_0^l dy \int_0^l \frac{dx}{r} \qquad (7.18)$$

Fig. 7.6 Mutual inductance between two equal parallel filaments of length l against their separation d according to Eq. (7.21). The arrangement is shown in Fig. 7.5

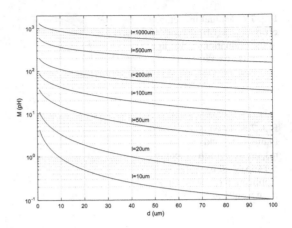

x, y in (7.18) run along the lengths of filaments f_1 and f_2, respectively. Setting $r = \sqrt{(y-x)^2 + d^2}$ and evaluating the inner integral result in

$$M = \frac{\mu_o}{4\pi} \int_0^l \left[\text{arcsinh}\left(\frac{y}{d}\right) - \text{arcsinh}\left(\frac{y-l}{d}\right) \right] dy \qquad (7.19)$$

The integral in (7.19) gives

$$M = \frac{\mu_o}{2\pi} l \left[\text{arcsinh}\left(\frac{l}{d}\right) + \frac{d}{l} - \sqrt{1 + \frac{d^2}{l^2}} \right] \qquad (7.20)$$

Finally, using $\text{arcsinh}(x) = ln(x + \sqrt{1 + x^2})$ transforms (7.20) into [28]

$$M = \frac{\mu_o}{2\pi} l \left[ln\left(\frac{l}{d} + \sqrt{1 + \frac{l^2}{d^2}}\right) + \frac{d}{l} - \sqrt{1 + \frac{d^2}{l^2}} \right] \qquad (7.21)$$

Figure 7.6 shows the mutual inductance between the two equal parallel filaments depicted in Fig. 7.5 for different values of length l and separation d.

7.4 Mutual Inductance of Rectangular Parallel Bars

Let us now calculate the mutual inductance between two parallel bars with rectangular cross-sections, having equal length l and thickness t. The bar edges are separated by distance s, as is shown in Fig. 7.7. The widths of the bars are w_1 and w_2, respectively. For bars with cross-sectional dimensions small compared to their separation, it suffices to treat them as two filaments placed at their geometrical

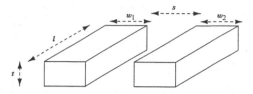

Fig. 7.7 Two parallel bars of rectangular cross-section, having equal length l and thickness t, separated by edge-to-edge distance s. The widths of the bars are w_1 and w_2, respectively

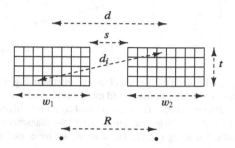

Fig. 7.8 Cross-sectional view of the two rectangular bars subdivided into filamentary sub-conductors. The center-to-center separation is $d = s + w_1/2 + w_2/2$. Their mutual inductance is approximately equal to the mutual between two filaments placed apart by the geometric-mean distance R of the two areas that represent the cross-sections of the segments

centers. For bars too close to justify this approximation, it is necessary to break them up into filamentary sub-conductors and average the mutual inductances of all the filaments [28, 30]. This is shown in Fig. 7.8. The approximation is valid for low frequencies where the current density is uniform over the bars' cross-sectional areas.

Figure 7.9 shows the mutual inductance between two bars depicted in Fig. 7.7, using two different methods. Both bars have equal length $l = 100\,\mu\text{m}$, widths $w_1 = w_2 = 20\,\mu\text{m}$, and thickness $t = 3.3\,\mu\text{m}$. The x-axis shows edge-to-edge separation s. The continuous curve is the mutual inductance calculated by averaging the mutual inductances of all the filaments using Eq. (7.21). The dashed curve is the mutual inductance calculated using (7.21), where for d we have used the center-to-center separation of the bars. The two methods give similar results as the edge-to-edge separation increases but deviate for small separations. We expect the deviation to be more significant as the bar widths increase.

An additional method to estimate the mutual between the two bars is utilizing the geometric-mean distance (GMD). The mutual inductance between the two bars is approximated by the mutual between two filaments placed apart by the geometric-mean distance R of the two areas that represent the cross-sections of the bars as is shown in the lower part of Fig. 7.8 [28, 30].

To see how the concept of GMD enters the picture, let us consider Eq. (7.21) in the limiting case where $l \gg d$. Discarding all the d/l terms gives

$$M \approx \frac{\mu_0}{2\pi} l \left[ln\left(\frac{2l}{d}\right) - 1 \right] \qquad (7.22)$$

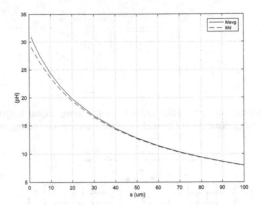

Fig. 7.9 Mutual inductance between two parallel bars as depicted in Fig. 7.7. Both bars have equal length $l = 100\,\mu$m, widths $w_1 = w_2 = 20\,\mu$m, and thickness $t = 3.3\,\mu$m. The x-axis shows edge-to-edge separation s. Continuous curve: The bars are broken up into filamentary sub-conductors, and the mutual inductance is estimated by averaging over all the filaments using Eq. (7.21). Dashed curve: The mutual is estimated using (7.21), where for d, we have used the bar center-to-center separation of the bars

Equation (7.22) can be further written as

$$M \approx \frac{\mu_o}{2\pi} l \left[ln(2l) - ln(d) - 1 \right]$$
(7.23)

Only the term $ln(d)$ in (7.23) depends on the filament separation. When calculating the average of the mutual inductances of the filamentary sub-conductors, we average the term ln(d), which results in

$$\frac{1}{M} \sum_{i=1}^{M} \ln(d_i) = ln \left[\left(\prod_{i=1}^{M} d_i \right)^{\frac{1}{M}} \right] = ln(R)$$
(7.24)

where terms d_i are the distances between any two filaments in Fig. 7.8. R in (7.24) is the geometric-mean distance of the areas that represent the cross-sections of the bars

$$R = \left(\prod_{i=1}^{M} d_i \right)^{\frac{1}{M}}$$
(7.25)

For computation, it is more convenient to use

$$R = e^{\frac{1}{M} \sum_{i=1}^{M} \ln(d_i)}$$
(7.26)

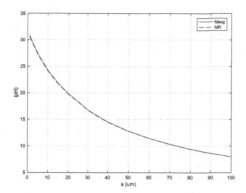

Fig. 7.10 Mutual inductance between two parallel bars depicted in Fig. 7.7. Both bars have equal length $l = 100\,\mu$m, widths $w_1 = w_2 = 20\,\mu$m, and thickness $t = 3.3\,\mu$m. The x-axis shows edge-to-edge separation s. Continuous curve: The bars are broken up into filamentary sub-conductors, and the mutual inductance is estimated by averaging over all the filaments using Eq. (7.21). Dashed curve: The mutual inductance between the bars is estimated using (7.21), where for d, we have used the geometric-mean distance R of the areas that represent the cross-sections of the bars

To get a better approximation of (7.21) for $l \gg d$, we expand the d/l terms in a series as is shown below

$$M \approx \frac{\mu_o}{2\pi} l \left[ln\left(\frac{2l}{d}\right) - 1 + \frac{d}{l} - \frac{d^2}{2l^2} + ... \right] \qquad (7.27)$$

Therefore, while the average of $ln(d)$ is the principal term, the average of d, d^2, etc., must also be considered [30]. Utilizing Eq. (7.21) with $d = R$ is therefore an approximation. It turns out that it is quite a good one, as is shown in Fig. 7.10, where the continuous curve shows the mutual inductance between the bars using the average method, while the dashed curve gives the mutual inductance between the bars using the geometric-mean method. Both bars have equal length $l = 100\,\mu$m, widths $w_1 = w_2 = 20\,\mu$m, and thickness $t = 3.3\,\mu$m. The x-axis shows edge-to-edge separation s. The two curves are almost on top of each other, even for small bar separations.

The continuous curve in Fig. 7.11 shows the error in estimating the mutual inductance when using the geometric-mean distance R in (7.21), in comparison to the averaging method. The dashed curve gives the error in the mutual inductance estimation when using the center-to-center separation d in (7.21), compared to the averaging method. The bars have equal length $l = 100\,\mu$m, widths $w_1 = w_2 = 20\,\mu$m, and thickness $t = 3.3\,\mu$m. The x-axis shows edge-to-edge separation s. The former method keeps the error small, even for small edge-to-edge separations.

The continuous curve in Fig. 7.12 shows the simulated mutual inductance between two rectangularly shaped bars, while the dashed curve gives the estimated mutual inductance using the geometric-mean distance R in (7.21). The two bars depicted in Figs. 7.7 and 7.8 have equal length $l = 100\,\mu$m, width $w_1 = w_2 =$

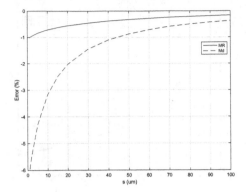

Fig. 7.11 Continuous curve: Error in the mutual inductance estimation when using the geometric-mean distance R in (7.21), in comparison to the averaging method. Dashed curve: Error in the mutual inductance estimation when using the center-to-center separation d in (7.21), compared to the averaging method. The bars have equal length $l = 100\,\mu m$, widths $w_1 = w_2 = 20\,\mu m$, and thickness $t = 3.3\,\mu m$. The x-axis shows edge-to-edge separation s

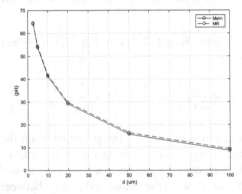

Fig. 7.12 Continuous curve: Simulated mutual inductance between two rectangularly shaped bars. Dashed curve: Estimated mutual inductance using the geometric-mean distance R in (7.21). The two bars depicted in Figs. 7.7 and 7.8 have equal length $l = 100\,\mu m$, width $w_1 = w_2 = 1.8\,\mu m$, and thickness $t = 3.3\,\mu m$. The x-axis shows center-to-center separation d

1.8 μm, and thickness $t = 3.3\,\mu m$. The x-axis shows center-to-center separation d. A very good agreement between simulation and estimation is observed.

7.5   Self-inductance of Rectangular Bars

The concept of geometric-mean distance is also useful when evaluating the self-inductance of a rectangularly shaped bar shown in Fig. 7.13. In Fig. 7.14, the bar is subdivided into filamentary sub-conductors, and its self-inductance is calculated

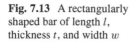

Fig. 7.13 A rectangularly shaped bar of length l, thickness t, and width w

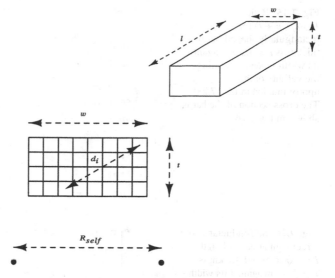

Fig. 7.14 Cross-sectional view of the rectangular bar shown in Fig. 7.13. To find the self-inductance of the bar, we subdivide it into filamentary sub-conductors and average the mutuals between all the filaments. d_i represents the distances between any two filaments. The self-inductance of the bar may also be evaluated by the mutual inductance between two filaments placed apart by the self-geometric-mean distance R_{self} of the area that represents the cross-section of the bar

by averaging the mutual inductances of all the filaments. Alternatively the self-inductance of the bar may be estimated by the mutual inductance between two filaments placed apart by the self-geometric-mean distance R_{self} of the area that represents the cross-section of the bar.

For rectangularly shaped bars, the self-geometric-mean distance can be approximated by the simple expression [28, 30]

$$R_{self} \approx 0.2235(w + t) \tag{7.28}$$

Figure 7.15 plots the ratio $R_{self}/(w + t)$, for a rectangularly shaped bar of thickness $t = 3.3\,\mu\text{m}$ against its width w, demonstrating the validity of the approximation in Eq. (7.28).

To obtain an expression for the self-inductance of a rectangular bar of length l, width w, and thickness t, we start from Eq. (7.21), which gives the mutual inductance between two filaments of length l, separated by distance d. We proceed with the reasonable assumption that $\frac{l^2}{d^2} \gg 1$, which simplifies (7.21) to

$$L \approx \frac{\mu_o}{2\pi} l \left[ln\left(\frac{2l}{d}\right) + \frac{d}{l} - 1 \right] \tag{7.29}$$

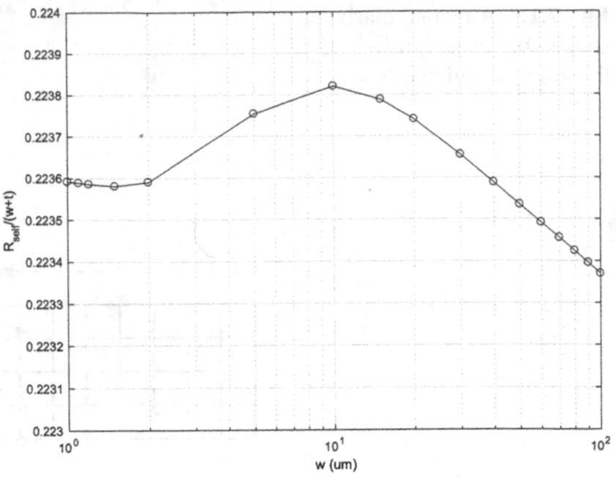

Fig. 7.15 Ratio $R_{self}/(w+t)$ for a rectangularly shaped bar of thickness $t = 3.3\,\mu\mathrm{m}$ against its width w, demonstrating the validity of the approximation in Eq. (7.28). The cross-section of the bar is shown in Fig. 7.14

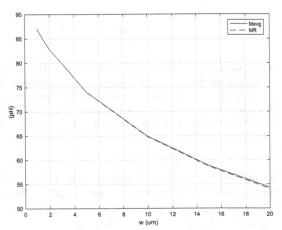

Fig. 7.16 Self-inductance of a rectangular bar of length $l = 100\,\mu\mathrm{m}$ and thickness $t = 3.3\,\mu\mathrm{m}$ against its width w. Continuous curve: The self-inductance is estimated using the averaging method. Dashed curve: The self-inductance is estimated using Eq. (7.30)

Substituting for d the expression for the self-geometric-mean distance R_{self} given in (7.28) results in [28, 30]

$$L \approx \frac{\mu_o}{2\pi} l \left[ln\left(\frac{2l}{w+t}\right) + \frac{1}{2} + 0.2235 \frac{w+t}{l} \right] \qquad (7.30)$$

where we have used that $ln(0.2235) \approx 1.5$.

The continuous curve in Fig. 7.16 shows the estimated self-inductance of a rectangular bar of length $l = 100\,\mu\mathrm{m}$ and thickness $t = 3.3\,\mu\mathrm{m}$ against its width w using the averaging method. The dashed curve corresponds to the case where the self-inductance is estimated using (7.30). The two curves are almost on top of each other. Figure 7.17 shows the error in estimating the self-inductance when using (7.30), in comparison to the averaging method. The error remains very small even for relatively large bar widths, despite the approximation $\frac{l^2}{d^2} \gg 1$ in deriving (7.30).

Fig. 7.17 Error in estimating the self-inductance of a rectangularly shaped bar when using (7.30), in comparison to the averaging method, against the bar width w. The bar has length $l = 100\,\mu\mathrm{m}$ and thickness $t = 3.3\,\mu\mathrm{m}$

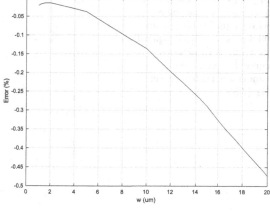

Fig. 7.18 Self-inductance of a rectangularly shaped bar with width $w = 1.8\,\mu\mathrm{m}$ and thickness $t = 3.3\,\mu\mathrm{m}$ against its length l. Continuous curve: Electromagnetic simulation. Dashed curve: Eq. (7.30)

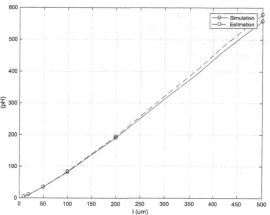

Figures 7.18 and 7.19 compare the self-inductance of a rectangular bar estimated using Eq. (7.30) with electromagnetic simulations. Good agreement is observed over considerable bar length and width variations.

7.6 Inductance of Rectangular Planar Loops

Rectangular planar inductors consist of N straight segments, as shown in Fig. 7.20. They are characterized by an array whose elements are the mutual inductances between the segments they consist of

$$
M = \begin{bmatrix}
M_{11} & M_{12} & \dots & M_{1N} \\
M_{21} & M_{22} & \dots & M_{2N} \\
\vdots & \vdots & \ddots & \vdots \\
M_{N1} & M_{N2} & \dots & M_{NN}
\end{bmatrix}
\tag{7.31}
$$

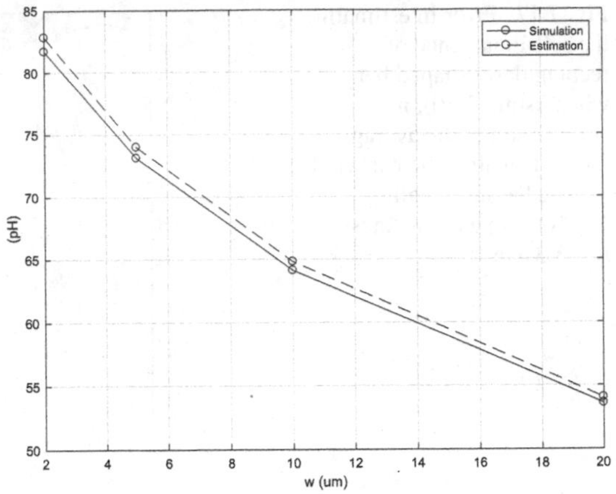

Fig. 7.19 Self-inductance of a rectangularly shaped bar with length $l = 100\,\mu$m and thickness $t = 3.3\,\mu$m against its width w. Continuous curve: Electromagnetic simulation. Dashed curve: Eq. (7.30)

Fig. 7.20 A two-turn rectangular planar inductor that consists of $N = 13$ straight segments

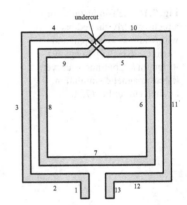

In (7.31), M_{ij} represents the mutual inductance between segments i and j, while M_{ii} denotes the self-inductance of a segment. To evaluate the total inductance of the structure, we sum the elements of the array

$$L = \sum_{i=1}^{N} \sum_{j=1}^{N} M_{ij} \tag{7.32}$$

Combining Neumann's form given in Eq. (7.17) with Eq. (7.32), we obtain for the inductance of a N-segment rectangular planar loop the following equation:

$$L = \sum_{i=1}^{N} \sum_{j=1}^{N} \left[\frac{\mu_o}{4\pi} \int d\mathbf{l}_j \int \frac{d\mathbf{l}_i}{r} \right] \tag{7.33}$$

In (7.33), segments i, j are replaced by filaments spaced apart by the corresponding geometric-mean distance R_{ij}. The term in the brackets evaluates Neumann's form on any two filaments.

Fig. 7.21 To estimate the total inductance of the rectangular planar inductor shown on the upper part of the figure, we replace all segments with filaments as is shown on the lower part. We subsequently apply Eq. (7.33) after placing the filaments apart by the geometric-mean distance R of the areas representing the corresponding segments' cross-sections

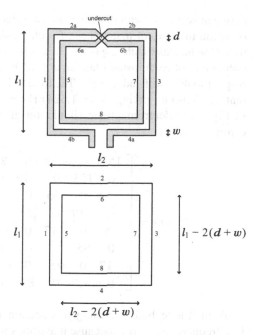

Filaments that are perpendicular to each other do not couple. This is because vectors $d\mathbf{A}$ and $d\mathbf{l}_B$ shown in Fig. 7.4 are perpendicular to each other. Similarly, filaments that run parallel to each other add to the total inductance if the current through them flows in the same direction (for example, segments 3 and 8 in Fig. 7.20) and subtract from the total inductance otherwise (for instance, segments 8 and 6 in Fig. 7.20). Filaments couple stronger the closer they are. Therefore, to maximize the total inductance, filaments with the same current direction must be kept close together, while filaments with opposite current directions must be kept further apart.

For example, we estimate the mutual inductance matrix of the rectangular planar inductor shown in the upper part of Fig. 7.21. Both turns have width $w = 10\,\mu\text{m}$, thickness $t = 3.3\,\mu\text{m}$ and are separated by distance $d = 5\,\mu\text{m}$. The outer dimensions of the planar inductor are $l_1 = l_2 = 200\,\mu\text{m}$. We replace all segments with filaments as is shown in the lower part of Fig. 7.21. We also merge segments $2a - 2b$, $4a - 4b$, and $6a - 6b$ into filaments 2, 4, and 6, respectively. As noted above, filaments perpendicular to each other do not couple. The mutual inductance between filaments that are parallel to each other is given by Eq. (7.17) after we separate the filaments by the geometric-mean distance of the areas that represent the cross-sections of the corresponding segments. The total inductance is given by Eq. (7.33) with $N = 8$.

The mutual inductance matrix is given in (7.34) in pH. Segments perpendicular to each other have zero mutual inductance. The self-inductance of the outer segments is 155 pH, while the self-inductance of the inner segments is 126 pH. The mutual inductance between adjacent segments is 85 pH. The mutual inductances between segments with opposite current directions are less than $-20\,\mathrm{pH}$. The contribution to the total inductance of each outside filament is $203\,\mathrm{pH}$, corresponding to a 31% increase with respect to their self-inductance. The contribution to the total inductance of each inside filament is $176\,\mathrm{pH}$, corresponding to a 40% increase with respect to their self-inductance. The total inductance is obtained by summing all the matrix elements, giving $L = 1.524\,\mathrm{nH}$. For comparison, for the planar inductor in Fig. 7.21, electromagnetic simulation gives $L = 1.529\,\mathrm{nH}$, very close to the estimated value.

$$M(pH) = \begin{bmatrix} 155 & 0 & -19 & 0 & 85 & 0 & -17 & 0 \\ 0 & 155 & 0 & -19 & 0 & 85 & 0 & -17 \\ -19 & 0 & 155 & 0 & -17 & 0 & 85 & 0 \\ 0 & -19 & 0 & 155 & 0 & -17 & 0 & 85 \\ 85 & 0 & -17 & 0 & 126 & 0 & -16 & 0 \\ 0 & 85 & 0 & -17 & 0 & 126 & 0 & -16 \\ -17 & 0 & 85 & 0 & -16 & 0 & 126 & 0 \\ 0 & -17 & 0 & 85 & 0 & -16 & 0 & 126 \end{bmatrix} \tag{7.34}$$

A final note before closing this section is that (7.33) applies strictly only at low frequencies. This is because it assumes uniform current density over the cross-sections of the inductor segments. Due to the skin and proximity effects, the current density ceases to be uniform at high frequencies. Such phenomena dramatically affect the segment resistance at high frequencies but generally only have a minor impact on the inductance value.

7.7   Skin Effect

The quality factor of a planar inductor is determined by energy loss in the inductor structure. A predominant loss mechanism at high frequencies is the skin effect, which is attributed to the tendency of the current in an isolated conductor to forsake the interior of the conductor's cross-section. Charge carriers prefer to crowd nearer to the conductor's surface, increasing its resistance with frequency. The presence of nearby conductors confines the allowed paths of charged carriers further, thus resulting in even higher resistance as frequency increases. This is known as the proximity effect. Displacement currents flow in the silicon substrate underneath the inductor structure due to capacitive coupling and dissipate energy in the finite silicon substrate resistance. This further exacerbates energy loss with frequency. This and the following sections will describe methods to characterize loss due to such effects.

To obtain an analytical approximation of the skin effect, we consider the attenuation of an electromagnetic wave as it propagates into a semi-infinite conductor characterized by magnetic permeability μ, dielectric constant ϵ, and conductivity σ as is depicted in Fig. 7.22. The field at the conductor's surface is the applied

Fig. 7.22 Electromagnetic
wave with components E_x
and B_y, propagating along the
z-direction, into a
semi-infinite conductor

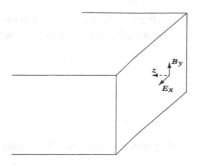

field counteracted by induced fields from time-varying currents underneath the conductor's surface. We start from Maxwell's equations [31].

$$\nabla \times \mathbf{E} = -\frac{\partial \mathbf{B}}{\partial t} \tag{7.35}$$

$$\nabla \times \mathbf{B} = \mu \mathbf{J} \tag{7.36}$$

where $\mathbf{J} = \sigma \mathbf{E}$ is the current density vector. Equation (7.35) is Faraday's law, while Eq. (7.36) is Ampere's law. We have ignored the displacement current in (7.36) since in a good conductor, the conductivity current is orders of magnitude larger than the displacement current ($\sigma \gg \epsilon\omega$). To solve (7.35) and (7.36) inside a semi-infinite conductor, we allow field variation only along the z-axis, which is the direction of propagation

$$\frac{\partial}{\partial x} = 0 \tag{7.37}$$

$$\frac{\partial}{\partial y} = 0 \tag{7.38}$$

$$\frac{\partial}{\partial z} \neq 0 \tag{7.39}$$

As is shown in Fig. 7.22, the electric field points to the x-direction, while the magnetic field points to the y-direction. Equations (7.35) and (7.36) become

$$\frac{\partial E_x}{\partial z} = -\frac{\partial B_y}{\partial t} \tag{7.40}$$

$$-\frac{\partial B_y}{\partial z} = \mu \sigma E_x \tag{7.41}$$

Decoupling the two equations, we obtain for the electric field a diffusion equation

$$\frac{\partial^2 E_x}{\partial z^2} = \mu \sigma \frac{\partial E_x}{\partial t} \tag{7.42}$$

To solve (7.42), we assume a wave solution of the form $E_x = E_o e^{-j(\omega t - kz)}$, where E_o is the field at the surface of the conductor and k is the wavenumber. Then

$$\frac{\partial}{\partial z} \equiv jk \tag{7.43}$$

$$\frac{\partial}{\partial t} \equiv -j\omega \tag{7.44}$$

Substituting (7.43) and (7.44) into (7.42), we obtain for the wavenumber

$$k^2 = j\mu\sigma\omega \tag{7.45}$$

which results in

$$k = k_r + jk_i = \sqrt{\frac{\mu\omega\sigma}{2}}(1+j) \tag{7.46}$$

The wavenumber is complex, with the imaginary part describing attenuation and the real part phase retardation in the direction of propagation. The current density J_x thus becomes

$$J_x = \sigma E_x = \sigma E_o e^{-k_i z} e^{-j\omega t} e^{jk_r z} \tag{7.47}$$

We use (7.46) to define the skin depth of the conductor as $\delta = \sqrt{2/\mu\sigma\omega}$. The physical interpretation of δ is that the magnitude of the current density J_x inside the conductor decays to 37% of its value at the surface $J_o = \sigma E_o$ at depth equal to the skin depth. The skin depth of copper at room temperature is about 1 μm at 5 GHz. The total current per unit width produced by the field is

$$I = \int_0^\infty \sigma E_o e^{\frac{-z}{\delta}} dz = \sigma E_o \delta \tag{7.48}$$

We define the surface resistivity as the resistance per unit length per unit width by

$$R_s \equiv \frac{E_o}{I} = \frac{1}{\sigma\delta} \tag{7.49}$$

Effectively, the current flows underneath the surface, inside a layer of depth equal to the skin depth, and the resistance of the conductor is proportional to the square root of the frequency. Equation (7.49) describes accurately the ac-resistance of a conductor only in cases where the conductor depth and the radius of curvature are much larger than the skin depth. For inductor segments with cross-sectional dimensions not greater than a few micrometers, it is only a crude approximation, and an alternative approach is required.

7.8 Segment Resistance at High Frequencies

The high-frequency resistance of a segment can be accurately modeled by subdividing it into N coupled, filamentary sub-conductors as is shown in Figs. 7.23 and 7.24 [29]. Due to the mutual coupling, inner filaments have greater reactance than filaments closer to the surface, and internal current flow is deterred progressively as frequency increases. The net effect increases the segment's resistance with frequency as the current is confined to a smaller cross-sectional area. This kind of loss is thus attributed solely to magnetic effects. The sub-conductor model assumes that the dominant electric field component is along the segment's length and treats segment cross-sections as equipotential surfaces. This is reasonably accurate away from the segment's ends.

Each sub-conductor is modeled by its dc-resistance R in series with its self-inductance L. Mutual coupling between all the sub-conductors is also included. We can demonstrate that such a model is theoretically sound by substituting Eq. (7.9) into Faraday's law given in (7.35) [26]

$$\nabla \times \mathbf{E} = -\frac{\partial (\nabla \times \mathbf{A})}{\partial t} \tag{7.50}$$

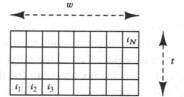

Fig. 7.23 Cross-sectional view of an isolated rectangular segment with width w and thickness t. The tendency of high-frequency currents to flow nearer to the surface is modeled by subdividing the segment into N coupled filamentary sub-conductors. The current in each sub-conductor is denoted by i_1, i_2, i_3, etc.

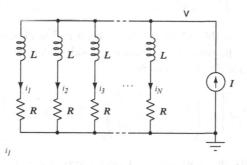

Fig. 7.24 Each sub-conductor is modeled by its dc-resistance R in series with its self-inductance L. Mutual coupling between all the sub-conductors is also taken into account

which becomes

$$\nabla \times \left(\mathbf{E} + \frac{\partial \mathbf{A}}{\partial t} \right) = 0 \tag{7.51}$$

In analogy to electrostatics, Eq. (7.51) invites the introduction of a scalar potential V according to

$$\mathbf{E} + \frac{\partial \mathbf{A}}{\partial t} = -\nabla V \tag{7.52}$$

Finally, substituting $\mathbf{J} = \sigma \mathbf{E}$, we obtain

$$-\nabla V = \frac{\mathbf{J}}{\sigma} + \frac{\partial \mathbf{A}}{\partial t} \tag{7.53}$$

Applying (7.53) along the length l of sub-conductor i with cross-sectional area S, we obtain

$$V_i = \frac{1}{\sigma} \frac{l}{S} i_i + L \frac{di_i}{dt} + \sum_{\substack{j=1 \\ j \neq i}}^{N} M_{ij} \frac{di_j}{dt} \tag{7.54}$$

where M_{ij} represents the mutual inductance between sub-conductors i and j. The first term in the sum corresponds to the voltage drop on the sub-conductor's dc-resistance, estimated by the segment's conductivity and the sub-conductor's geometrical dimensions. The second term accounts for the sub-conductor's self-inductance L, while the third term accounts for the mutual coupling between sub-conductor i and the rest of the sub-conductors in the segment. In order to estimate L and M_{ij} in (7.54), we use the results derived in Sects. 7.3 and 7.5 repeated below

$$L \approx \frac{\mu_o}{2\pi} l \left[ln \left(\frac{2l}{dw + dt} \right) + \frac{1}{2} + 0.2235 \frac{dw + dt}{l} \right] \tag{7.55}$$

where dw and dt are the sub-conductor's width and thickness, respectively. The mutual inductance between two filamentary sub-conductors of equal length l separated by distance d is estimated by

$$M = \frac{\mu_o}{2\pi} l \left[ln \left(\frac{l}{d} + \sqrt{1 + \frac{l^2}{d^2}} \right) + \frac{d}{l} - \sqrt{1 + \frac{d^2}{l^2}} \right] \tag{7.56}$$

For accurate results, the cross-sectional dimensions of the sub-conductors dw and dt must be kept smaller than the skin depth.

Fig. 7.25 Current density
($\mu A/\mu m^2$) over the
cross-section of an isolated
copper inductor segment with
width $w = 20\,\mu$m, thickness
$t = 3.3\,\mu$m, at 1 GHz, at 25C
($\sigma_{copper} = 5.2 \cdot 10^7$ S/m).
The ac-resistance is estimated
to be 363 Ω/m. A total
current of 1 mA is assumed to
run in the segment

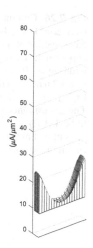

Assuming that we force current I into the segment (see Fig. 7.24), the individual
sub-conductor currents, along with the voltage V across the segment, can be
obtained by solving the system of equations

$$
\begin{bmatrix} 0 \\ 0 \\ \vdots \\ 0 \\ I \end{bmatrix} = \begin{bmatrix} R+sL & sM_{12} & \dots & sM_{1N} & -1 \\ sM_{21} & R+sL & \dots & sM_{2N} & -1 \\ \vdots & \vdots & \ddots & \vdots & \vdots \\ sM_{N1} & sM_{N2} & \dots & R+sL & -1 \\ 1 & 1 & \dots & 1 & 0 \end{bmatrix} \cdot \begin{bmatrix} i_1 \\ i_2 \\ \vdots \\ i_N \\ V \end{bmatrix} \tag{7.57}
$$

The segment's high-frequency resistance can be approximated by

$$
R_{ac} = \frac{real(V)}{I} \tag{7.58}
$$

Figures 7.25, 7.26, 7.27, and 7.28 show the current density over the cross-section
of an isolated copper inductor segment with cross-sectional dimensions $w = 20\,\mu$m
and $t = 3.3\,\mu$m, at 25C, at 1 GHz, 2 GHz, 5 GHz, and 10 GHz, respectively. The
copper conductivity at 25C is $\sigma_{copper} = 5.2 \cdot 10^7$ S/m. The ac-resistance of the
segment rises fast from 363 Ω/m at 1 GHz to 431 Ω/m at 2 GHz, to 607 Ω/m at
5 GHz, and to 854 Ω/m at 10 GHz. The dc-resistance of the segment at 25C is
291 Ω/m. The current density increases steeply as frequency rises, especially near
the corners of the cross-section of the conductor. In all three cases, a total current of
1 mA is assumed to run in the segment's cross-section. For the accurate computation
of the high-frequency resistance, the sub-conductors' cross-sectional dimensions are
smaller than the skin depth in each case.

Fig. 7.26 Current density $(\mu A/\mu m^2)$ over the cross-section of an isolated copper inductor segment with width $w = 20\,\mu m$, thickness $t = 3.3\,\mu m$, at 2 GHz, at 25C $(\sigma_{copper} = 5.2 \cdot 10^7\,\text{S/m})$. The ac-resistance is estimated to be 431 Ω/m. A total current of 1 mA is assumed to run in the segment

Fig. 7.27 Current density $(\mu A/\mu m^2)$ over the cross-section of an isolated copper inductor segment with width $w = 20\,\mu m$, thickness $t = 3.3\,\mu m$, at 5 GHz, at 25C $(\sigma_{copper} = 5.2 \cdot 10^7\,\text{S/m})$. The ac-resistance is estimated to be 607 Ω/m. A total current of 1 mA is assumed to run in the segment

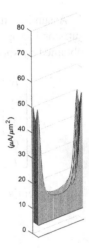

The complex current distribution over the segment's cross-section confirms that attempting to predict high-frequency resistance using the skin-depth approximation would only result in a crude estimation. Figure 7.29 depicts the estimated ac-resistance in $k\Omega/m$ of an isolated copper inductor segment with thickness $t = 3.3\,\mu m$ versus width and frequency at room temperature.

The continuous curve in Fig. 7.30 shows the simulated ac-resistance of an isolated copper inductor segment with width $w = 20\,\mu m$ and thickness $t = 3.3\,\mu m$, against frequency, at 25C. The dashed curve shows the estimated ac-resistance using the method of sub-conductors developed in this section. Good agreement between simulated and calculated results is observed, with the worst discrepancy being about 7%.

Fig. 7.28 Current density ($\mu A/\mu m^2$) over the cross-section of an isolated copper inductor segment with width $w = 20\,\mu m$, thickness $t = 3.3\,\mu m$, at 10 GHz, at 25C ($\sigma_{copper} = 5.2 \cdot 10^7$ S/m). The ac-resistance is estimated to be 854 Ω/m. A total current of 1 mA is assumed to run in the segment

Fig. 7.29 Estimation of the ac-resistance in $k\Omega/m$ of an isolated copper inductor segment with thickness $t = 3.3\,\mu m$, against its width and frequency, at 25C ($\sigma_{copper} = 5.2 \cdot 10^7$ S/m) with the method of sub-conductors

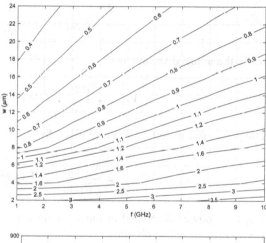

Fig. 7.30 High-frequency resistance in Ω/m of an isolated copper inductor segment with width $w = 20\,\mu m$ and thickness $t = 3.3\,\mu m$, against frequency, at 25C ($\sigma_{copper} = 5.2 \cdot 10^7$ S/m). Continuous curve: Electromagnetic simulation. Dashed curve: Estimation with the method of sub-conductors

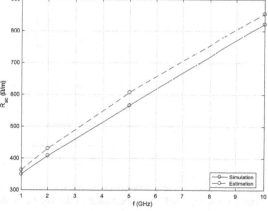

7.9   Proximity Effect

The mutual coupling between the conductor under investigation and nearby conductors further confines the current flow, resulting in even higher resistance with
frequency. The phenomenon can be treated by the method of sub-conductors.
Still, in the specific case of small conductivity, a simple analysis can reveal the
dependence on frequency and segment dimensions [25].

Let us assume a conductive segment with width w, length l, and thickness h as is
depicted in Fig. 7.31. We further assume that the magnetic field inside the segment is
independent of position and is determined solely by the externally applied field. This
is a reasonable approximation if the segment's conductivity is sufficiently low. The
time variation of the external magnetic field $B = B_o \sin(\omega t)$ results in Eddy currents
inside the segment. The resistance of the Eddy current path shown in Fig. 7.31 can
be approximated by

$$R_{path} = \frac{1}{\sigma} \frac{2l}{h \, dx} \tag{7.59}$$

while the induced voltage is

$$V_{path} = \frac{d\Phi}{dt} = (2lx) \, B_o \, \omega \cos(\omega t) \tag{7.60}$$

The power dissipated is therefore given by

$$P_{path} = \frac{V^2_{path_{rms}}}{R_{path}} = \left(\sigma l B_o^2 \, \omega^2 h\right) x^2 dx \tag{7.61}$$

The total power dissipated in the segment is estimated by integrating (7.61) from
$x = 0$ to $x = w/2$ to obtain

$$P_{total} = \frac{\sigma l h}{24} B_o^2 \omega^2 w^3 \tag{7.62}$$

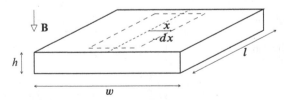

Fig. 7.31 Loss inside a low-conductivity segment due to a position-independent, time-varying
external magnetic field is proportional to the square of the frequency and the third power of the
segment width

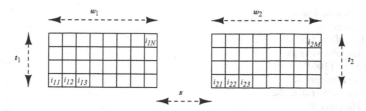

Fig. 7.32 Cross-sectional view of two parallel inductor segments. The proximity effect between the two segments can be modeled by subdividing them into N and M coupled filamentary sub-conductors

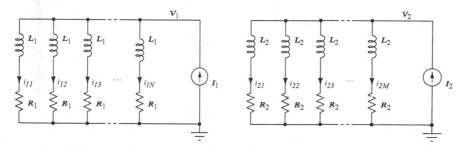

Fig. 7.33 Each sub-conductor is modeled by its dc-resistance and its self-inductance. The mutual coupling between all sub-conductors is also considered

Equation (7.62) shows a strong dependence on the segment's width and frequency. One way to reduce the strong dependence on the width is to break up the segment along its width into N smaller pieces. Following a similar analysis, the total power dissipated goes down by N^2. This technique can reduce the power dissipated in the shield layer underneath the planar inductor structure.

Given the assumptions in deriving (7.62), one can only expect this result to be a crude approximation for a practical inductor segment. An alternative way is the method of sub-conductors. Let us assume two inductor segments separated by distance s as shown in Fig. 7.32. The segment on the left (segment 1) has cross-sectional dimensions w_1 and t_1, while the segment on the right (segment 2) has cross-sectional dimensions w_2 and t_2. We subdivide each segment into N and M coupled sub-conductors, respectively. Each sub-conductor is modeled by its dc-resistance and its self-inductance. To account for the mutual coupling between the two segments, each sub-conductor on the left segment is coupled with all the sub-conductors in that segment and all the sub-conductors in the segment on the right and vice versa.

Figure 7.33 shows an equivalent circuit model. Assuming that we force current I_1 into segment 1 and current I_2 into segment 2, the sub-conductor currents and the voltages V_1 and V_2 across the two segments can be obtained by solving the system of equations given in (7.63).

Fig. 7.34 Current density
($\mu A/\mu m^2$) over the
cross-section of the left
segment. The ac-resistance is
estimated to be $1106\,\Omega/\text{m}$ at
$5\,\text{GHz}$. A total current of
$1\,\text{mA}$ is assumed to run in the
segment. The current
distribution is shifted toward
the left, away from segment 2
that is positioned on the right

$$
\begin{bmatrix} 0 \\ \vdots \\ 0 \\ 0 \\ \vdots \\ 0 \\ I_1 \\ I_2 \end{bmatrix} = \begin{bmatrix} R_1 + sL_1 & \cdots & sM_{11,1N} & sM_{11,21} & \cdots & sM_{11,2M} & -1 & 0 \\ \vdots & \ddots & \vdots & \vdots & \ddots & \vdots & \vdots & \vdots \\ sM_{1N,11} & \cdots & R_1 + sL_1 & sM_{1N,21} & \cdots & sM_{1N,2M} & -1 & 0 \\ sM_{21,11} & \cdots & sM_{21,1N} & R_2 + sL_2 & \cdots & sM_{21,2M} & 0 & -1 \\ \vdots & \ddots & \vdots & \vdots & \ddots & \vdots & \vdots & \vdots \\ sM_{2M,11} & \cdots & sM_{2M,1N} & sM_{2M,21} & \cdots & R_2 + sL_2 & 0 & -1 \\ 1 & \cdots & 1 & 0 & \cdots & 0 & 0 & 0 \\ 0 & \cdots & 0 & 1 & \cdots & 1 & 0 & 0 \end{bmatrix} \cdot \begin{bmatrix} i_{11} \\ \vdots \\ i_{1N} \\ i_{21} \\ \vdots \\ i_{2M} \\ V_1 \\ V_2 \end{bmatrix}
$$
$$(7.63)$$

In Eq. (7.63), $M_{1i,1j}$ is the mutual inductance between sub-conductors i and j in segment 1. $M_{1i,2j}$ is the mutual inductance between sub-conductor i in segment 1 and sub-conductor j in segment 2. $M_{2i,2j}$ is the mutual inductance between sub-conductors i and j in segment 2. $M_{2i,1j}$ is the mutual inductance between sub-conductor i in segment 2 and sub-conductor j in segment 1. All the sub-conductors in segment 1 are modeled by a resistance R_1 in series with inductance L_1. Similarly, all the sub-conductors in segment 2 are modeled by a resistor R_2 in series with inductance L_2. These are estimates of the dc-resistance and self-inductance of the filaments. Finally, current i_{1i} represents the current flowing in the sub-conductor i of segment 1, while current i_{2i} represents the current flowing in the sub-conductor i of segment 2.

Figures 7.34 and 7.35 show the current density over the cross-section of the two segments for $w_1 = w_2 = 10\,\mu\text{m}$, $t_1 = t_2 = 3.3\,\mu\text{m}$, at room temperature, at $5\,\text{GHz}$. The segments are separated by distance $d = 5\,\mu\text{m}$. The ac-resistance of the two

Fig. 7.35 Current density
$(\mu A/\mu m^2)$ over the
cross-section of the right
segment. The ac-resistance is
estimated to be 1106 Ω/m at
5 GHz. A total current of
1 mA is assumed to run in the
segment. The current
distribution is shifted toward
the right, away from segment
1 positioned on the left

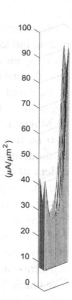

Fig. 7.36 High-frequency
resistance at 25C of two
segments with thickness
$t = 3.3\,\mu$m and width
$w = 10\,\mu$m, separated by
distance $d = 5\,\mu$m.
Continuous curve:
Electromagnetic simulation.
Dashed curve: Estimation
using the method of
sub-conductors

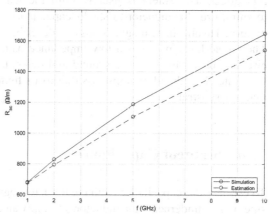

segments is estimated to be about 1106 Ω/m. The segments' ac-resistance due to
the skin effect alone is 1031 Ω/m at room temperature at 5 GHz. We have assumed
that the current in both segments flows in the same direction, repelling the current
distributions as shown in Figs. 7.34 and 7.35. A total current of 1 mA is assumed to
run in the segments' cross-sections.

The continuous curve in Fig. 7.36 shows the simulated high-frequency resistance
at 25C of two copper inductor segments with thickness $t = 3.3\,\mu$m and width
$w = 10\,\mu$m against frequency. The segments are separated by distance $d = 5\,\mu$m.
The dashed curve shows the estimated ac-resistance using the method of sub-
conductors. Good agreement between simulated and calculated results is observed,
with the worst discrepancy being less than 7%.

The methodology can be extended to more than two adjacent segments, as in planar inductor structures with more than two turns. As mentioned in Sect. 7.6, bringing inductor segments with the same current direction closer together increases their mutual coupling and, thus, the overall inductance of the structure. The tradeoff is increased ac-resistance due to proximity, resulting in higher energy loss. Larger segment width results in smaller ac-resistance but also larger segment capacitance.

Magnetic loss in the silicon substrate underneath the inductor structure can also be tackled with the method of sub-conductors by setting $I_2 = 0$ in (7.63). In modern silicon processes, the silicon substrate conductance $\sigma_{sub} \approx 10\,\mathrm{S/m}$ is orders of magnitude lower than the thick-metal conductance $\sigma_{copper} \approx 10^7\,\mathrm{S/m}$, and the magnetic loss in the silicon substrate is very small compared to the loss in the inductor turns. Effectively, the Eddy currents generated in the substrate are negligible.

On the other hand, electric loss in the substrate becomes significant as frequency increases. Displacement currents are injected into the substrate via the segment-substrate coupling capacitance and dissipate energy in the substrate's finite resistance. We utilize a shield layer underneath the inductor structure to mitigate this effect. The shield is usually implemented with the lowest metal available to minimize the inductor-to-shield capacitance. Furthermore, the shield layer is patterned to eliminate magnetic loss, as described in the opening part of this section. The shield layer provides a low impedance path for the displacement currents to circulate or terminate to the ground and deters them from escaping into the lossy substrate. The shield exhibits loss due to its finite resistance, albeit much smaller than the substrate.

7.10 Segment Capacitance

To estimate the capacitance from the inductor segment to the silicon substrate or the shield layer underneath the inductor, we apply an external voltage V on the segment and compute the total charge Q_{seg} deposited on it. The silicon substrate or shield is kept at zero volts. The segment capacitance is given by

$$C_{seg} = \frac{Q_{seg}}{V} \qquad (7.64)$$

As the substrate relaxation time $\epsilon_{sub}/\sigma_{sub} \approx 10\,psec$ is much smaller than the signal period for frequencies below 10 GHz, we may treat the substrate surface as an ideal ground plane. This is also valid for a high-conductivity shield layer underneath the planar inductor.

To calculate the segment charge Q_{seg} per meter of length, we model it as N line-charge distributions, each with line-charge density λ_i as depicted in Fig. 7.37. The induced charge on the silicon substrate or shield layer is denoted by Q_{sur}.

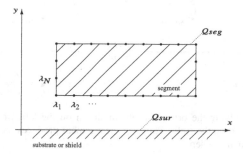

Fig. 7.37 Segment cross-sectional view above the silicon substrate or shield layer. The length of the segment is perpendicular to the plane of the page. The segment charge Q_{seg} per meter of length is approximated by N line-charge distributions, each with a line-charge density λ_i. The induced charge on the silicon substrate or shield layer is denoted by Q_{sur}

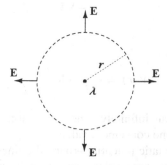

Fig. 7.38 The electric field around an infinitely long line-charge distribution can be estimated by applying the Maxwell–Gauss law on the surface of a cylinder that symmetrically encloses the charge distribution. Because of the radial symmetry, the electric field is perpendicular to the cylinder surface, and its value depends only on r. The length of the line charge runs perpendicularly to the plane of the page

To proceed, we first calculate the electrostatic potential around a single line-charge distribution in the absence of a ground plane, using the Maxwell–Gauss law

$$\nabla \cdot \mathbf{E} = \frac{Q_{enc}}{\epsilon} \qquad (7.65)$$

where Q_{enc} is the charge enclosed in the area on which we calculate the electric field. The dielectric constant ϵ is an average value estimated by the dielectric constants of the inter-metal dielectric layers of the process. Applying (7.65) on the surface of a cylinder of length l that symmetrically encloses the line-charge distribution as is shown in Fig. 7.38 gives the magnitude of the electric field

$$E\,(2\pi r\, l) = \frac{\lambda\, l}{\epsilon} \Rightarrow E = \frac{\lambda}{2\pi \epsilon\, r} \qquad (7.66)$$

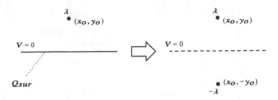

Fig. 7.39 Instead of solving the original problem shown on the left, we study the completely different situation shown on the right. This new arrangement produces the same potential above the surface as the original problem

The electric field vector points radially outward in Fig. 7.38 for positive λ and inward otherwise. The electrostatic potential V at every point in space is defined by

$$\mathbf{E} = -\nabla V \tag{7.67}$$

giving

$$V(r) = -\frac{\lambda}{2\pi\epsilon} \ln\left(\frac{r}{r_o}\right) \tag{7.68}$$

As we are dealing with an infinitely long line-charge distribution, we have set $V(r_o) = 0$, where r_o is some convenient value.

To estimate the electrostatic potential around the line-charge distribution in the presence of the ground plane, we need to account for the contribution to the potential of the charge Q_{sur} that is attracted on the ground plane. The solution to this problem requires using the uniqueness theorem, which states that the potential in a region is uniquely determined by the charge density throughout the region and the value of the potential at its boundaries [26].

As is shown in Fig. 7.39, instead of studying the original problem shown on the left, we study the completely different situation shown on the right. In this new problem, we discard the silicon substrate and its surface charge. Instead, we place an image line-charge distribution with charge density $-\lambda$ at a distance y_o below the line defined by the silicon substrate surface. Due to the symmetry, the potential on the line is zero. The only charge above the line is the same as in the original problem. By the uniqueness theorem, the new arrangement produces the same potential as the original arrangement in the region above the line. This approach is known as the method of images [25, 26]. The potential at any point (x, y) above the line in Fig. 7.39 is, therefore, the sum of the potential due to the original line-charge distribution plus the potential of the image distribution. This consideration results in

$$V(x, y) = \frac{\lambda}{2\pi\epsilon} \ln \sqrt{\frac{(x - x_o)^2 + (y + y_o)^2}{(x - x_o)^2 + (y - y_o)^2}}, \quad y \geq 0 \tag{7.69}$$

Returning to the original problem depicted in Fig. 7.37, to evaluate the electrostatic potential at any point above the silicon substrate surface, we need to account for the contributions of all the line-charge distributions $\lambda_i(x_i, y_i)$, as well as, for the contributions of the associated image distributions $-\lambda_i(x_i, -y_i)$. This results in

$$V_{tot}(x, y) = \frac{1}{2\pi\epsilon} \sum_{i=1}^{N} \lambda(x_i, y_i) \ln \sqrt{\frac{(x - x_i)^2 + (y + y_i)^2}{(x - x_i)^2 + (y - y_i)^2}} \qquad (7.70)$$

The line-charge densities λ_i can now be determined if we compute (7.70) on the surface of the segment, where the potential is constant and equal to the externally applied voltage V. This results in the system of equations

$$\begin{bmatrix} V \\ V \\ \vdots \\ V \end{bmatrix} = \frac{1}{2\pi\epsilon} \cdot \begin{bmatrix} g_{1,1} & g_{1,2} & \cdots & g_{1,N} \\ g_{2,1} & g_{2,2} & \cdots & g_{2,N} \\ \vdots & \vdots & \vdots & \vdots \\ g_{N,1} & g_{N,2} & \cdots & g_{N,N} \end{bmatrix} \cdot \begin{bmatrix} \lambda_1 \\ \lambda_2 \\ \vdots \\ \lambda_N \end{bmatrix} \qquad (7.71)$$

Terms $g_{j,i}$ are given by

$$g_{j,i} = \ln \sqrt{\frac{(x_j - x_i)^2 + (y_j + y_i)^2}{(x_j - x_i)^2 + (y_j - y_i)^2}}, \quad i \neq j \qquad (7.72)$$

In (7.72), the line-charge distributions λ_i are located at (x_i, y_i), and the potential is estimated at (x_j, y_j). To avoid the infinities in the diagonal terms in (7.71), we compute the potentials at points that are very slightly offset from (x_i, y_i). Solving for λ_i in (7.71) allows us to estimate the segment capacitance by simply summing the computed line-charge densities

$$C_{seg} = l \cdot \sum_{i=1}^{N} \lambda_i \qquad (7.73)$$

where l is the length of the segment. The method just outlined is a simple form of a technique known as the method of moments [25, 32].

Computing the potential $V(x, y)$ just above $y = 0$ in Fig. 7.37 allows us to also estimate the surface charge density σ_{sur} as

$$\sigma_{sur}(x) = \epsilon \frac{\partial V(x, y)}{\partial y}\bigg|_{y=0} \qquad (7.74)$$

Figure 7.40 shows the induced σ_{sur} on the silicon substrate or shield layer underneath an inductor segment. The segment has width $w = 10\,\mu m$ and thickness $t = 3.3\,\mu m$ and is $3.525\,\mu m$ above the substrate or shield. The method of moments

Fig. 7.40 Induced charge density σ on the silicon substrate/shield underneath an inductor segment with width $w = 10\,\mu$m and thickness $t = 3.3\,\mu$m. The segment is maintained at 1 V. With reference to Fig. 7.37, the segment's lower left corner is located at $x = 0$, $y = 3.485\,\mu$m

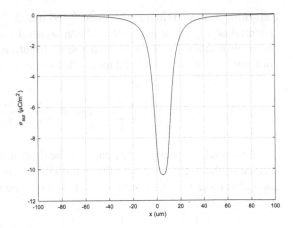

gives a segment capacitance of 22.1 fF per 100 μm length, in close agreement with the simulated value 22.5 fF. For the capacitance estimation, we have used $\epsilon = 4.15\epsilon_o$, where the permittivity of free space is $\epsilon_o = 8.85$ pF/m.

In the case of multi-turn inductors, we are interested in estimating the capacitances from each segment to the substrate or shield and the segment-to-segment capacitances. It is straightforward to extend the method of moments to cover this case. As an example, for two 100 μm long inductor segments of equal width $w = 10\,\mu$m and thickness $t = 3.3\,\mu$m, separated by distance $s = 5\,\mu$m, we obtain 18.9 fF from each segment to the substrate or shield, and 4.7 fF between the two segments. These values are very close to the simulated results of 19.5 fF and 4.6 fF, respectively.

7.11   Inductor Equivalent Network

A simple equivalent π-network for a differential inductor is shown in Fig. 7.41. It captures the dominant effects that influence inductor performance and allows us to determine the dimensions of the inductor structure before resorting to electromagnetic simulations.

Inductor L captures the low-frequency inductance calculated with the method developed in Sect. 7.6. We replace all inductor segments with filaments and apply Eq. (7.33) after we space them apart by the geometric-mean distance of the areas that represent the cross-sections of the corresponding segments. Resistor r accounts for magnetic loss due to the skin and proximity effects. For single-turn inductors, the magnetic loss is dominated by the skin effect, while for multi-turn inductors, we also account for the proximity of the inductor turns. For processes with low substrate conductivity, the magnetic loss in the substrate is minimal and can be disregarded. By applying the method of sub-conductors developed in Sects. 7.8 and 7.9, we

Fig. 7.41 Equivalent
network for a planar
differential inductor

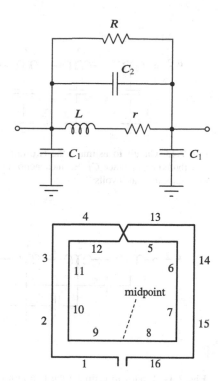

Fig. 7.42 Two-turn
differential inductor broken
up into 16 segments to
capture the distributed nature
of the segment-to-shield C_{sh}
and segment-to-segment C_c
capacitors

estimate the magnetic loss series resistance r_i for each inductor segment. The value
of r is given by the sum

$$r = \sum_{i=1}^{N} r_i \qquad (7.75)$$

Magnetic loss increases with frequency; therefore, r is frequency dependent.

The values of the segment-to-shield C_{sh} and segment-to-segment C_c capacitances are estimated with the method of moments developed in Sect. 7.10. The
method estimates capacitance per length, so to obtain C_{sh} and C_c, we need to
multiply with the perimeters of the inductor turns. Capacitors C_{sh} and C_c are
distributed. As an example, in the case of the differential two-turn inductor depicted
in Fig. 7.42, we account for the distributed nature of C_{sh} and C_c using the equivalent
circuits shown in Figs. 7.43 and 7.44.

With reference to Fig. 7.43, and assuming differential drive, the energy stored in
the sixteen capacitors is given by [8]

$$E = 2 \cdot \left[\frac{1}{2} \left(\frac{C_{sh}}{16} \right) \sum_{i=1}^{8} \left(\frac{i}{16} v \right)^2 \right] = 2 \cdot \left[\frac{1}{2} \left(\frac{C_{sh}}{5} \right) \left(\frac{v}{2} \right)^2 \right] \qquad (7.76)$$

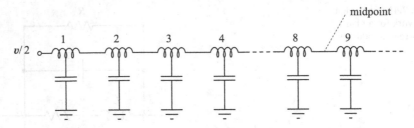

Fig. 7.43 Circuit to estimate the two-turn differential inductor's π-network segment-to-shield distributed capacitance C_1. Each capacitor in the figure is given by $C_{sh}/16$. For differential drive, the midpoint is at 0 volts

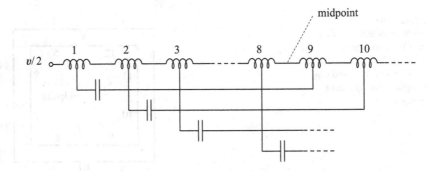

Fig. 7.44 Circuit to estimate the two-turn differential inductor's π-network distributed capacitance C_2. Each capacitor in the figure is given by $C_c/8$. The midpoint is at 0 volts for a differential drive, and each capacitor sees voltage $v/2$ across it

The value of the distributed segment-to-shield capacitor C_1 in Fig. 7.41 is therefore

$$C_1 = \frac{C_{sh}}{5} \tag{7.77}$$

Similarly, with reference to Fig. 7.44, and assuming differential drive, the energy stored in the eight capacitors is given by [8]

$$E = 8 \cdot \left[\frac{1}{2} \left(\frac{C_c}{8} \right) \left(\frac{v}{2} \right)^2 \right] = \frac{1}{2} \left(\frac{C_c}{4} \right) v^2 \tag{7.78}$$

The value of the distributed segment-to-segment capacitor C_2 in Fig. 7.41 is therefore

$$C_2 = \frac{C_c}{4} \tag{7.79}$$

Resistor R in Fig. 7.41 approximates the loss in the finite resistance of the shield underneath the inductor structure. Figure 7.45 depicts a typical floating shield for

Fig. 7.45 Two-turn
differential inductor with
floating metal shield. The
shield is implemented with
the lowest metal available to
minimize the
inductor-to-shield
capacitance. Furthermore, the
shield layer is patterned to
eliminate magnetic loss

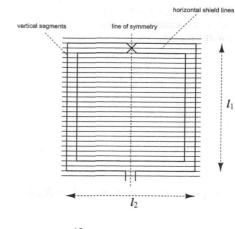

Fig. 7.46 Cross-sectional view of the vertical inductor segments. C_s is the capacitance from
a vertical segment of the inductor to the horizontal shield lines. Resistance R_s is the overall
resistance of the horizontal shield lines from underneath the inductor's vertical segment to the
line of symmetry that cuts the structure in two, as is shown in Fig. 7.45. The midpoint is a virtual
ground for differential driving $\pm v/2$

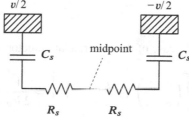

a differential inductor. As mentioned, the shield is implemented with the lowest
metal available to minimize the inductor-to-shield capacitance and is patterned to
eliminate magnetic loss.

In Fig. 7.45, the floating shield lines provide a low impedance path for the
displacement currents emanating from the inductor segments and deter them from
flowing into the lossy substrate. For differential driving, the points of the shield lines
intersecting the line of symmetry are at zero potential. The shield exhibits loss due
to its finite resistance, albeit smaller than the substrate.

In Fig. 7.46, C_s is the capacitance from a vertical segment of the inductor to
the horizontal shield lines. Resistance R_s is the overall resistance of the horizontal
shield lines from underneath the inductor's vertical segment to the line of symmetry
that cuts the structure in two, as is shown in Fig. 7.45.

In Fig. 7.47, we transform the series $C_s - R_s$ network into a parallel one [7]. The
value of the parallel resistor R_p is given by

$$R_p = R_s \left(1 + \frac{1}{R_s^2 \, \omega^2 \, C_s^2} \right) \tag{7.80}$$

Fig. 7.47 The series $C_s - R_s$ network is transformed into a parallel $C_p//R_p$ one

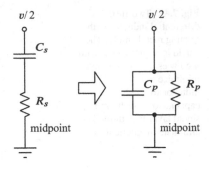

The power dissipated on resistor R_p is

$$P_{R_p} = \frac{(v/2)^2}{2\,R_p} \tag{7.81}$$

Applying Eq. (7.81) to the case of the differential inductor shown in Fig. 7.45 gives

$$P_{shield} = 2\,\frac{1}{2\,R_p}\left[\left(\frac{6}{8}\cdot\frac{v}{2}\right)^2 + \left(\frac{2}{8}\cdot\frac{v}{2}\right)^2\right] = \frac{1}{2\,R_p}\left(\frac{5}{16}\right)v^2 \tag{7.82}$$

In deriving (7.82), we took into account that the voltages in the middle of the external and internal vertical segments of the two-turn differential inductor are $\frac{6}{8}(\frac{v}{2})$ and $\frac{2}{8}(\frac{v}{2})$, respectively. From Eqs. (7.82) and (7.80), the value of the shield loss resistor R in Fig. 7.41 becomes

$$R = \frac{16}{5}R_p = \frac{16}{5}R_s\left(1 + \frac{1}{R_s^2\,\omega^2\,C_s^2}\right) \tag{7.83}$$

Like r, the value of R is also frequency dependent. The value of R_s is

$$R_s = \frac{100}{d}\,\rho\,\frac{l_2}{2\,l_1} \tag{7.84}$$

l_1 and l_2 are the inductor dimensions depicted in Fig. 7.45, while ρ and d are the shield layer resistance per square and fill density in percentage, respectively.

For example, let us calculate the equivalent network for the rectangular planar inductor shown in Fig. 7.21. The turns have width $w = 10\,\mu m$, thickness $t = 3.3\,\mu m$, and are separated by distance $d = 5\,\mu m$. The structure is positioned $3.071\,\mu m$ above the shield layer, which has resistance $\rho = 0.647$ Ohm /square and 50% fill density. The outer dimensions of the planar inductor are $l_1 = l_2 = 200\,\mu m$.

In Sect. 7.6, we estimated the dc inductance to be $L = 1.524\,nH$. Since this is a two-turn inductor on top of a shield layer, skin and proximity effects dominate

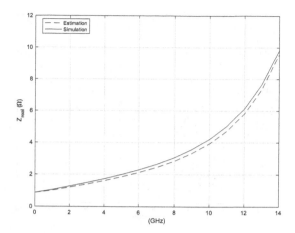

Fig. 7.48 Real part of the two-turn rectangular planar inductor impedance shown in Fig. 7.21. The turns have width $w = 10\,\mu$m, thickness $t = 3.3\,\mu$m, and are separated by distance $d = 5\,\mu$m. The structure is positioned $3.071\,\mu$m above the shield layer, which has resistance $\rho = 0.647$ Ohm /square and 50% fill density. The outer dimensions of the planar inductor are $l_1 = l_2 = 200\,\mu$m. Continuous curve: Electromagnetic simulation. Dashed curve: Estimation

the magnetic loss. Applying the method of sub-conductors presented in Sects. 7.8 and 7.9, we obtain $r = 1.0\,\Omega$, $r = 1.2\,\Omega$, $r = 1.6\,\Omega$, and $r = 2.3\,\Omega$, at 1 GHz, 2 GHz, 5 GHz, and 10 GHz, respectively. The method of moments in Sect. 7.10 gives $C_{sh} = 306.8$ fF and $C_c = 31.3$ fF for the segment-to-shield and the segment-to-segment capacitances. Subsequently, Eqs. (7.77) and (7.79) give for the distributed equivalents $C_1 = 61.4$ fF and $C_2 = 7.8$ fF. Finally, setting $C_s = C_{sh}/4$ in Eqs. (7.83) and (7.84), we obtain $R = 21.3\,\text{M}\Omega$, $R = 5.3\,\text{M}\Omega$, $R = 852\,\text{k}\Omega$, and $R = 213\,\text{k}\Omega$, at 1 GHz, 2 GHz, 5 GHz, and 10 GHz, respectively.

Figures 7.48 and 7.49 compare the estimated and simulated real and imaginary parts of the inductor impedance. Similarly, Figs. 7.50 and 7.51 compare the estimated and simulated inductance and quality factor. The estimated real part of the impedance agrees with the simulation over a wide frequency range. The imaginary part also agrees with simulation for frequencies up to 10 GHz and deviates more as frequency increases. This is because of the quasistatic approximations adopted in developing the equivalent network. This is also reflected in the estimated inductance, which deviates from simulation as frequency increases and exhibits about 12% discrepancy at 14 GHz. The estimated quality factor shows a good prediction of the peak-Q frequency and varies less than 15% from the simulation in that region.

In this chapter, we have discussed the analysis and design of planar spiral inductors for fully integrated VCOs in the GHz range. Starting from the first principles, we gradually built a π-model that encapsulates the essential effects that determine inductor performance, such as inductance value, parasitic capacitances, and magnetic and electric loss. The model agrees with electromagnetic simulations

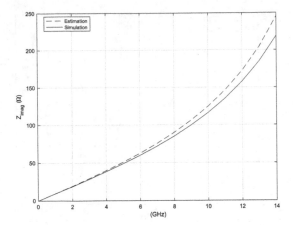

Fig. 7.49 Imaginary part of the two-turn rectangular planar inductor impedance shown in Fig. 7.21. The turns have width $w = 10\,\mu$m, thickness $t = 3.3\,\mu$m, and are separated by distance $d = 5\,\mu$m. The structure is positioned $3.071\,\mu$m above the shield layer, which has resistance $\rho = 0.647$ Ohm /square and 50% fill density. The outer dimensions of the planar inductor are $l_1 = l_2 = 200\,\mu$m. Continuous curve: Electromagnetic simulation. Dashed curve: Estimation

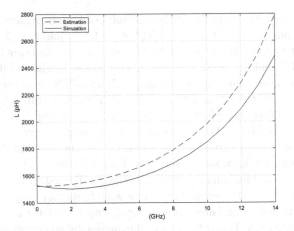

Fig. 7.50 Inductance of the two-turn rectangular planar inductor shown in Fig. 7.21. The turns have width $w = 10\,\mu$m, thickness $t = 3.3\,\mu$m, and are separated by distance $d = 5\,\mu$m. The structure is positioned $3.071\,\mu$m above the shield layer, which has resistance $\rho = 0.647$ Ohm /square and 50% fill density. The outer dimensions of the planar inductor are $l_1 = l_2 = 200\,\mu$m. Continuous curve: Electromagnetic simulation. Dashed curve: Estimation

over a wide frequency range and helps obtain an initial estimation of the inductor dimensions before resorting to electromagnetic simulations for final optimization. It also highlights the limitations of quasistatic analysis methods as frequency increases.

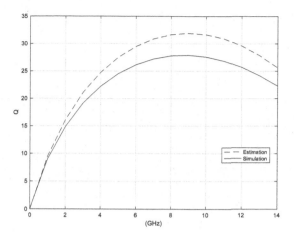

Fig. 7.51 Quality factor of the two-turn rectangular planar inductor shown in Fig. 7.21. The turns have width $w = 10\,\mu\text{m}$, thickness $t = 3.3\,\mu\text{m}$, and are separated by distance $d = 5\,\mu\text{m}$. The structure is positioned $3.071\,\mu\text{m}$ above the shield layer, which has resistance $\rho = 0.647\,\text{Ohm /square}$ and 50% fill density. The outer dimensions of the planar inductor are $l_1 = l_2 = 200\,\mu\text{m}$. Continuous curve: Electromagnetic simulation. Dashed curve: Estimation

Appendix A
Amplitude Noise

A.1 Stochastic Differential Equation for Amplitude

In the idealized limit cycle shown in Fig. 3.5, charge Δq injected into the tank during time Δt induces a voltage change $\Delta q / C$ in the tank capacitor. The amplitude deviation r is calculated as

$$r = \Delta A \approx \frac{\Delta q}{C} \sin(\omega_o t) \qquad (A.1)$$

As noted, the amplitude is most sensitive to noise fluctuations when the carrier reaches its peak values. It has been assumed that such amplitude fluctuations are suppressed by the nonlinearity of the self-sustained oscillator with infinite speed and do not contribute to the oscillator linewidth broadening. To obtain a more realistic model of the time evolution of amplitude fluctuations, we surmise that their decay is governed by the same mechanism that starts up the oscillator and stabilizes its amplitude. Thus, it is reasonable to assume that for small amplitude deviations around the oscillator limit cycle, the decay speed is proportional to the tank capacitance C and inversely proportional to $g_1 - 1/R$. A dynamical equation that for small amplitude deviations around the limit cycle expresses the competition mechanism between linear damping and random fluctuations is

$$\dot{r}_t = -\beta_r r_t + \frac{1}{C} \sin(\omega_o t) \, n_t \qquad (A.2)$$

The subscript t denotes time dependency, while overdot stands for time derivative. n_t is the fluctuating current across the tank. The value of the damping rate is approximated as

$$\beta_r = \frac{g_1 - 1/R}{C} = \frac{1}{R_r C} \qquad (A.3)$$

© The Author(s), under exclusive license to Springer Nature Switzerland AG 2023
K. Manetakis, *Topics in LC Oscillators*,
https://doi.org/10.1007/978-3-031-31086-7

Equation (A.2) models the behavior of small amplitude deviations around the limit cycle and can be regarded as the result of linearization around the equilibrium value that the oscillation amplitude attains on the limit cycle. It describes an Ornstein-Uhlenbeck process [16] and differs from (3.72) due to the appearance of the damping term $-\beta_r r_t$. This term expresses a restoring mechanism proportional to the magnitude of amplitude deviations. Equation (A.2) is further expressed in the form

$$dr_t = -\beta_r\, r_t\, dt + \frac{1}{C}\,\sin(\omega_o t)\, dn_t \tag{A.4}$$

where dn_t is a Wiener process [16]

$$dn_t = n_{t+dt} - n_t = \sqrt{\delta^2\, dt}\cdot N_t^{t+dt}(0,1) \tag{A.5}$$

and

$$\delta^2 = S_{i_n}(f) = \frac{\overline{i_n^2}}{df} = \frac{2kT}{R} \tag{A.6}$$

$N_t^{t+dt}(0,1)$ denotes a unit normal with zero mean and unit standard deviation. To solve the stochastic differential equation (A.4), we need to estimate the mean and the variance of r_t. For this, it is convenient to write it in the form

$$dr_t = -\beta_r\, r_t\, dt + \frac{\delta}{C}\,\sin(\omega_o t)\, dW_t \tag{A.7}$$

where $dW_t = \sqrt{dt}\cdot N_t^{t+dt}(0,1)$ is the unit Wiener process with zero mean and unit standard deviation [16]. Taking the average of (A.7) gives

$$d(\overline{r_t}) = -\beta_r\, \overline{r_t}\, dt \tag{A.8}$$

which with initial condition $r(0)$ results in

$$\overline{r_t} = r(0)\, e^{-\beta_r t} \tag{A.9}$$

Equation (A.9) confirms that an adequately small external disturbance unsettles r_t temporarily; it then returns to the equilibrium value attained on the limit cycle with a characteristic time $1/\beta_r$.

To estimate the variance of r_t, we start from

$$d\left(r_t^2\right) = (r_t + dr_t)^2 - r_t^2 \tag{A.10}$$

which results in

$$d\left(r_t^2\right) = 2\,r_t\,dr_t + (dr_t)^2 \tag{A.11}$$

Substituting (A.7) into (A.11) gives

$$d\left(r_t^2\right) = -2\,r_t^2\,\beta_r\,dt + 2\,r_t\frac{\delta}{C}\,\sin(\omega_o t)\,dW_t + \frac{\delta^2}{C^2}\sin^2(\omega_o t)\,(dW_t)^2 \tag{A.12}$$

Taking the average over the oscillator cycle and noting that $\overline{dW_t} = 0$, $\overline{\sin^2(\omega_o t)} = 1/2$ and $\overline{(dW_t)^2} = dt$ gives

$$d(\overline{r_t^2}) = \left(-2\,\beta_r\,\overline{r_t^2} + \frac{\delta^2}{2\,C^2}\right)dt \tag{A.13}$$

The solution of (A.13) for zero initial conditions is

$$\overline{r_t^2} = \frac{\delta^2}{4\,\beta_r\,C^2}(1 - e^{-2\,\beta_r\,t}) \tag{A.14}$$

The variance σ_r^2 of r_t is given by

$$\sigma_r^2 = \overline{r_t^2} - \overline{r_t}^2 \tag{A.15}$$

With zero initial conditions $\overline{r_t} = 0$. Therefore, (A.15) becomes

$$\sigma_r^2 = \overline{r_t^2} = \frac{\delta^2}{4\,\beta_r\,C^2}(1 - e^{-2\,\beta_r\,t}) \tag{A.16}$$

For times much larger than $1/2\beta_r$, the competition between random fluctuations and linear damping results in the equilibrium value

$$\sigma_r^2\bigg|_{t\gg 1/2\beta_r} = \frac{\delta^2}{4\,\beta_r\,C^2} = \frac{1}{2}\frac{k\,T}{C}\frac{1}{(g_1 R - 1)} \tag{A.17}$$

which shows that the amplitude noise power reduces for large tank capacitance C and large $g_1 R$ product. Large C results in small amplitude deviations for a given noise charge injected across the tank. Large $g_1 R - 1$ results in tighter amplitude regulation per the Van der Pol oscillator discussion in Chap. 2.

Being the outcome of a large number of uncorrelated random variables, we expect that the probability density function of r_t is Gaussian

$$P(r, t) = \frac{1}{\sqrt{2\pi\sigma_r^2}}e^{-\frac{(r_t - \overline{r_t})^2}{2\sigma_r^2}} \tag{A.18}$$

Fig. A.1 Time evolution of
amplitude fluctuations of a
self-sustained oscillator with
$C = 1\,\text{pF}, L = 1\,\text{nH},$
$R = 500$, and $g_1 = 3\,\text{mS}$

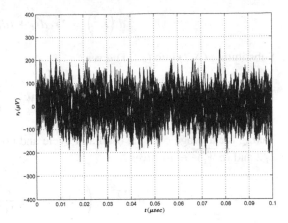

Fig. A.2 The probability
density for the amplitude
fluctuations of a
self-sustained oscillator with
$C = 1\,\text{pF}, L = 1\,\text{nH},$
$R = 500$ and $g_1 = 3\,\text{mS}$ for
$t \gg 1/2\beta_r$

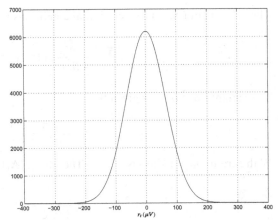

For times much larger than $1/2\beta_r$ Eq. (A.18) becomes independent of the time

$$P(r) = \frac{1}{\sqrt{2\pi \frac{\delta^2}{4\beta_r C^2}}} e^{-\frac{r^2}{2\frac{\delta^2}{4\beta_r C^2}}} \tag{A.19}$$

Figure A.1 shows ten different runs of the time evolution of the random amplitude.
The amplitude fluctuations do not diffuse with time, as is the case of phase
fluctuations depicted in Fig. 3.8. Figure A.2 shows the probability density of the
amplitude fluctuations for $t \gg 1/2\beta_r$. In both cases we have set $C = 1\,\text{pF}$,
$L = 1\,\text{nH}, R = 500$ and $g_1 = 3\,\text{mS}$.

A.2   Amplitude Noise

The oscillator carrier with amplitude noise is expressed as

$$V = (A_o + r_t)\,\sin(\omega_o t) \tag{A.20}$$

and its autocorrelation function

$$R_{VV}(t, t + \tau) = \overline{V(t)\,V(t + \tau)} \tag{A.21}$$

This is written as

$$R_{VV}(t, t + \tau) = \overline{(A_o + r_t)\sin(\omega_o t) \cdot (A_o + r_{t+\tau})\sin(\omega_o t + \omega_o \tau)} \tag{A.22}$$

Expanding the product in (A.22) gives

$$R_{VV}(t, t + \tau) = \frac{1}{2}\overline{(A_o + r_t)\,(A_o + r_{t+\tau}) \cdot (\cos(\omega_o \tau) - \cos(2\omega_o t + \omega_o \tau))} \tag{A.23}$$

The average value of the $\cos(2\omega_o t + \omega_o \tau)$ is zero. Similarly $\overline{r_t} = \overline{r_{t+\tau}} = 0$. Noting further that the term $\cos(\omega_o \tau)$ is constant, (A.23) becomes

$$R_{VV}(t, t + \tau) = \frac{1}{2}(A_o{}^2 + \overline{r_t\,r_{t+\tau}}) \cdot \cos(\omega_o \tau) \tag{A.24}$$

In order to calculate $\overline{r_t\,r_{t+\tau}}$, we start from (A.2) [19]

$$\frac{dr_{t+\tau}}{d\tau} = -\beta_r\,r_{t+\tau} + x_{t+\tau} \tag{A.25}$$

where $x_{t+\tau}$ is the noisy term. We subsequently multiply both sides by r_t to get

$$r_t\,\frac{dr_{t+\tau}}{d\tau} = -\beta_r\,r_t\,r_{t+\tau} + r_t\,x_{t+\tau} \tag{A.26}$$

Since r_t does not depend on τ, (A.26) can be expressed as

$$\frac{d(r_t\,r_{t+\tau})}{d\tau} = -\beta_r\,r_t\,r_{t+\tau} + r_t\,x_{t+\tau} \tag{A.27}$$

Taking the average of (A.27) results in

$$\frac{d\overline{(r_t\,r_{t+\tau})}}{d\tau} = -\beta_r\,\overline{r_t\,r_{t+\tau}} + \overline{r_t\,x_{t+\tau}} \tag{A.28}$$

Finally, since r_t and $x_{t+\tau}$ are uncorrelated, the second term on the right is zero, and (A.28) becomes

$$\frac{d\overline{(r_t\, r_{t+\tau})}}{d\tau} = -\beta_r\, \overline{r_t\, r_{t+\tau}} \tag{A.29}$$

The solution of (A.29) gives

$$\overline{r_t\, r_{t+\tau}} = \overline{r_t^{\,2}} \cdot e^{-\beta_r\, |\tau|} \tag{A.30}$$

We have introduced the absolute value since autocorrelation is an even function. Substituting (A.30) into (A.24) and using (A.17) give

$$R_{VV}(\tau) = \frac{1}{2} A_o^{\,2}\, \cos(\omega_o \tau) + \frac{\delta^2}{8\,\beta_r\, C^2} e^{-\beta_r\, |\tau|} \cos(\omega_o\, \tau) \tag{A.31}$$

The first term corresponds to the oscillator carrier, while the second captures amplitude noise fluctuations. With the aid of the Fourier transform properties (3.57), (3.58) and (3.59) we obtain the amplitude noise spectrum as

$$S_r(\omega) = \frac{\delta^2}{8\,C^2} \cdot \left[\frac{1}{(\omega_o + \omega)^2 + \beta_r^{\,2}} + \frac{1}{(\omega_o - \omega)^2 + \beta_r^{\,2}} \right] \tag{A.32}$$

The first term in the sum corresponds to negative frequencies, while the second term corresponds to the positive part of the spectrum. Equation (A.32) is further written as

$$S_r(\Delta\omega) = \frac{\delta^2}{8\,C^2} \cdot \frac{1}{(\Delta\omega)^2 + \beta_r^{\,2}} \tag{A.33}$$

where only positive frequencies are considered. $\Delta\omega = \omega_o - \omega$ is the frequency offset from the carrier. Equation (A.33) shows that amplitude noise exhibits no frequency dependence for values of $\Delta\omega$ up to about β_r and reduces beyond that. For frequency offsets adequately smaller than β_r, (A.33) can be simplified to

$$S_r|_{\Delta\omega \ll \beta_r} = S_r = \frac{\delta^2}{8\,C^2} \cdot \frac{1}{\beta_r^{\,2}} \tag{A.34}$$

As in the case of phase noise, we compare the amplitude noise spectral density (per Hz) at a frequency offset $\Delta\omega$ from the carrier to the carrier power. The amplitude noise-to-signal ratio can thus be written as

$$AN = \frac{S_r}{A_o^{\,2}/4} = \frac{\delta^2}{2\,C^2\, \beta_r^{\,2}\, A_o^{\,2}} = \frac{kTR}{A_o^{\,2}} \frac{1}{(g_1 R - 1)^2} \tag{A.35}$$

Fig. A.3 Phase and
amplitude noise plot for the
self-sustained oscillator
depicted in Fig. 3.4, with
$C = 1\,\text{pF}$, $L = 1\,\text{nH}$,
$R_p = 500\,\Omega$, $A_o = 1\,\text{V}$,
$g_1 = 5\,\text{mS}$. The x-axis shows
the frequency offset from the
carrier frequency. Circles:
Simulated Phase Noise.
Continuous line: Estimated
Phase Noise according to
(3.97). Triangles: Simulated
Amplitude Noise. Dashed
line: Estimated Amplitude
Noise according to (A.35)

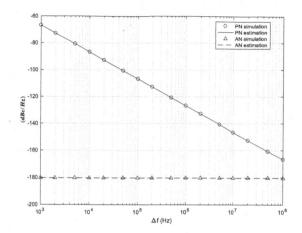

where AN is shorthand for amplitude noise. Figure A.3 compares simulation and Eq. (A.35) for the self-sustained oscillator shown in Fig. 3.4. We have used $L = 1\,\text{nH}$, $C = 1\,\text{pF}$, $R = 500$, $A_o = 1\,\text{V}$, and $g_1 = 5\,\text{mS}$. Excellent agreement between simulation and estimation is observed. The figure also depicts phase noise, demonstrating that phase noise is the dominant contributor to oscillator linewidth broadening for frequencies close to the carrier.

Appendix B
Derivation of Entrainment Equation

To solve (6.8) we rewrite it below as

$$\frac{d\phi_t}{dt} = \alpha \cdot \sin(\phi_t - \Delta\omega \cdot t) \tag{B.1}$$

where $\phi_t = \phi(t)$ and

$$\alpha = \frac{i_{nj_o}}{2A_1 C} \tag{B.2}$$

We subsequently set $\psi_t = \phi_t - \Delta\omega \cdot t$, which transforms (B.1) to

$$\frac{d\psi_t}{dt} = -\Delta\omega + \alpha \cdot \sin(\psi_t) \tag{B.3}$$

Integration of (B.3) gives [33]

$$\psi_t = 2\arctan\left[\frac{\alpha}{\Delta\omega} - \frac{\sqrt{\Delta\omega^2 - \alpha^2}}{\Delta\omega} \cdot \tan\left(\frac{\sqrt{\Delta\omega^2 - \alpha^2}}{2} \cdot t + C\right)\right] \tag{B.4}$$

where constant C satisfies the initial condition $\phi_t(0) = 0$. We can now express ϕ_t as

$$\phi_t = \Delta\omega \cdot t - 2\arctan\left[\frac{-\alpha}{\Delta\omega} + \frac{\sqrt{\Delta\omega^2 - \alpha^2}}{\Delta\omega} \cdot \tan\left(\frac{\sqrt{\Delta\omega^2 - \alpha^2}}{2} \cdot t + C\right)\right] \tag{B.5}$$

Setting $\phi_t(0) = 0$ gives for C

K. Manetakis, *Topics in LC Oscillators*,
https://doi.org/10.1007/978-3-031-31086-7

$$C = \arctan\left(\frac{\alpha}{\sqrt{\Delta\omega^2 - \alpha^2}}\right) \tag{B.6}$$

Combining (B.5) and (B.6) gives

$$\phi(t) = \Delta\omega \cdot t - 2\arctan\left[\frac{-\alpha + \sqrt{\Delta\omega^2 - \alpha^2} \cdot \tan\left[\frac{1}{2}\sqrt{\Delta\omega^2 - \alpha^2} \cdot t - \arctan\left(\frac{-\alpha}{\sqrt{\Delta\omega^2 - \alpha^2}}\right)\right]}{\Delta\omega}\right]$$

$$\tag{B.7}$$

which is (6.11). Equation (6.10) is derived from (B.7) by using the identities $\tan(jx) = j\tanh(x)$ and $\arctan(jx) = j\,\mathrm{arctanh}(x)$ [33].

References

1. Pain, H.J.: The Physics of Vibrations and Waves, 6th edn. Wiley, New York (2005)
2. Edwards, H.C., Penney, D.E.: Calculus, Early Transcendentals. Pearson Higher Education, London (2013)
3. Jordan, D., Smith, P.: Mathematical Techniques: An Introduction for the Engineering, Physical, and Mathematical Sciences. Oxford University, Oxford (2008)
4. Pikovsky, A., Kurths, J., Rosenblum, M., Kurths, J.: Synchronization: A Universal Concept in Nonlinear Science. Cambridge University, Cambridge (2001)
5. Kleppner, D., Kolenkow, R.: An Introduction to Mechanics. Cambridge University, Cambridge (2014)
6. Razavi, B.: Design of Analog CMOS Integrated Circuits. McGraw Hill, New York (2005)
7. Lee, T.H.: The Design of CMOS Radio-Frequency Integrated Circuits. Cambridge University, Cambridge (2003)
8. Razavi, B.: RF Microelectronics. Prentice Hall, Englewood (2012)
9. Acheson D.: From Calculus to Chaos: An Introduction to Dynamics. Oxford University, Oxford (1997)
10. Logan, J.D.: Applied Mathematics. Wiley, New York (2013)
11. Strogatz, S.H.: Nonlinear Dynamics and Chaos: With Applications to Physics, Biology, Chemistry, and Engineering. CRC Press, New York (2015)
12. Jordan, D., Smith, P.: Nonlinear Ordinary Differential Equations: An Introduction for Scientists and Engineers. Oxford University, Oxford (2007)
13. Manetakis, K.: Suppression of 1/f noise upconversion in LC-tuned oscillators. Electron. Lett. **50**(18), 1277–1279 (2014)
14. Tsividis, Y.: Mixed Analog-Digital VLSI Devices and Technology. World Scientific, Singapore (2002)
15. Reif, F.: Statistical Physics (Berkeley Physics Course—Volume 5). McGraw Hill, New York (1971)
16. Lemons, D.S.: An Introduction to Stochastic Processes in Physics. The Johns Hopkins University Press, Baltimore (2002)
17. MacDonald, D.: Noise and Fluctuations: an Introduction. Dover Publications, New York (2006)
18. Lesson, D.B.: A simple model of feedback oscillator noise spectrum. Proc. IEEE **54**(2), 329–330 (1966)
19. Dill, K.A., Bromberg, S., Stigter, D.: Molecular Driving Forces: Statistical Thermodynamics in Biology, Chemistry, Physics, and Nanoscience. Garland Science, New York (2010)
20. Lathi, B.P.: Modern Digital and Analog Communication Systems. Oxford University, Oxford (1995)

21. Gardiner, C.W.: Handbook of Stochastic Methods. Springer, Berlin (2003)
22. Hajimiri, A., Lee, T.H.: The Design of Low Noise Oscillators. Springer, Berlin (1999)
23. Adler, R.: A study of locking phenomena in oscillators. Proc. IRE **34**(6), 351–357 (1946)
24. Purcell, E.M., Morin, D.J.: Electricity and Magnetism. Cambridge University, Cambridge (2013)
25. Schwarz, S.E.: Electromagnetics for Engineers. Oxford University, Oxford (1990)
26. Griffiths, D.J.: Introduction to Electrodynamics. Pearson, London (2005)
27. Schey, H.M.: Div, Grad, Curl, and all that. W W Norton & Company, New York (1973)
28. Grover, F.W.: Inductance Calculations: Working Formulas and Tables. Dover, New York (1946)
29. Silvester, P.: Modal network theory of skin effect in flat conductors. Proc. IEEE **54**(9), 1147–1151 (1966)
30. Rosa, E.B.: The self and mutual inductances of linear conductors. In: US Department of Commerce and Labor, Bureau of Standards (80) (1908)
31. Solymar, L., Walsh, D., Syms, R.R.A.: Electrical Properties of Materials. Oxford University, Oxford (2014)
32. Sadiku, M.N.O.: Computational Electromagnetics with MATLAB. CRC Press, New York (2018)
33. Spiegel, M.R., Lipschutz, S., Liu, J.: Schaum's Outlines: Mathematical Handbook of Formulas and Tables. McGraw-Hill, London (2009)

Index

A

Adler's equation, 108
Ampère's law, 112, 125
Amplitude limiting, 14
Amplitude noise and phase noise comparison, 169
Amplitude noise for self-sustained oscillator, 167
Amplitude regulation, 14
Autocorrelation function of the carrier and correlation time of the fluctuations, 51
Autocorrelation function of the carrier and self-resemblance, 51
Autonomous system, 3

B

Barkhausen criteria, 13
Biot–Savart law, 109
Biot-Savart law, 124
Boltzmann distribution, 40

C

Competition between linear damping and random fluctuations, 163
Condition for oscillation, 14
Current distribution over segment's cross section, 144
Cyclostationary factor, 69

D

Damped harmonic oscillator, 4
Degrees of freedom, 40
Describing function approximation, 76

Deterministic part of phase, 53
Differential equation of damped harmonic motion, 5
Differential equation of simple harmonic motion, 2
Diffusion constant, 56
Diffusion equation, 49
Diffusion equation and stochastic differential equation, 56
Diffusion process, 48
Distributed capacitance, 155
Distributed segment-to-segment capacitor, 156
Distributed segment-to-shield capacitor, 156
Divergence theorem, 125
Driven harmonic oscillator, 8
Driven harmonic oscillator energy, 12

E

Effect of common mode on low-frequency noise to phase noise conversion, 85
Electric loss in the silicon substrate, 150
Energy balance, 22
Energy in the damped harmonic oscillator, 7
Energy restoring element, 13
Entrainment of self sustained oscillator, 99
Equilibrium between linear damping and random fluctuations, 165
Equipartition theorem, 40, 43
Excess noise factor, 70

F

Faraday's law, 123
fig8-shaped loop, 114

Printed in the United States
by Baker & Taylor Publisher Services